ACPL ITEM DISCARDED

693.
Santilli, Chris.
Opportunities in masonry careers

**DO NOT REMOVE
CARDS FROM POCKET**

**ALLEN COUNTY PUBLIC LIBRARY
FORT WAYNE, INDIANA 46802**

You may return this book to any agency, branch, or bookmobile of the Allen County Public Library.

VGM Opportunities Series

OPPORTUNITIES IN
MASONRY CAREERS

Chris Santilli

VGM Career Horizons
a division of *NTC Publishing Group*
Lincolnwood, Illinois USA

Cover Photo Credits:

Front cover: upper left, Oakton Community College; upper right, Brick Institute; lower left, *Aberdeen's Magazine of Masonry Construction;* lower right, courtesy of Roger Sheldon.

Back cover: upper left, courtesy of Chris Santilli; upper right and lower left, *Aberdeen's Magazine of Masonry Construction.*

Allen County Public Library
900 Webster Street
PO Box 2270
Fort Wayne, IN 46801-2270

Library of Congress Cataloging-in-Publication Data

Santilli, Chris
    Opportunities in masonry careers / Chris Santilli.
      p.   cm. — (VGM opportunities series)
      ISBN 0-8442-4066-4 — ISBN 0-8442-4067-2
      1. Masonry—Vocational guidance. I. Title. II. Series.
    TH5321.S26   1993
    693' .1' 023—dc20   92-37382
                                                                            CIP

Published by VGM Career Horizons, a division of NTC Publishing Group.
© 1993 by NTC Publishing Group, 4255 West Touhy Avenue,
Lincolnwood (Chicago), Illinois 60646-1975 U.S.A.
All rights reserved. No part of this book may be reproduced, stored
in a retrieval system, or transmitted in any form or by any means,
electronic, mechanical, photocopying, recording or otherwise, without
the prior permission of NTC Publishing Group.
Manufactured in the United States of America.

3 4 5 6 7 8 9 0 VP 9 8 7 6 5 4 3 2 1

ABOUT THE AUTHOR

For five years, Chris Santilli was the managing editor of various construction industry magazines such as *The Magazine of Masonry Construction* and *Concrete Construction*. She has ten years of experience as an editor, writer, and photographer. Currently she is free-lancing.

A native of Chicago, Ms. Santilli is a journalism graduate of Northern Illinois University and earned her master's in magazine journalism from Syracuse University.

FOREWORD

As trades continue to diminish in number during this Computer Age, there is still and always will be a need for masons. The skill and precision involved in masonry can never be replaced by machines, and the careful craftsmanship required to succeed as a mason ensures a steady need for young, creative minds to enter the field.

Look around you. There are very few places within civilization where you cannot see some form of masonry: a brick building, a stone wall, a cobbled drive. As artistic as they are functional, these works of craftsmanship are all products of the mason.

Workers through the ages have enjoyed seeing their efforts put to use and marveled at the product's ability to last a few years. Masonry, however, holds the distinction of being one of few trades whose work has withstood the stress and wear of centuries. Ancient engineering accomplishments such as the Egyptian Pyramids and the Roman aqueducts were the product of masons, as were the first skyscrapers built in the twentieth century.

Today masonry is experiencing a renaissance. The sterile glass and steel high-rises will continue to be built, but there is a resurgence in the building and construction industry to return to the beautiful and sculptured brick and stone buildings that were a bastion of architectural achievement a mere century ago.

As you enter the world of masonry take notice of what and who have come before you. Ahead of you lies the progression of a tradition that will live on through your ideas and work. Accept the challenge that is given you, and take pride in your accomplishments.

> The Editors
> VGM Career Books

INTRODUCTION

Most people have these concerns when they are choosing a career:

- Income. Will I earn adequate money and be covered by good benefits? (Chapter Four)
- Growth. Will I be able to advance, in both income and personal challenges? (Chapter Three)
- Satisfaction. Will I like the work? (Chapter Five)
- Employment. Will a job be available to me for the next 40 years I'll be working? (Chapter Eight)
- Qualifications. What training do I need, how long does it take, how much does it cost, and where can I get it? (Chapters Ten and Eleven)

Masonry construction, or what is sometimes called the trowel trades, offers many advantages that most industries do not. It pays well, offers you many employment opportunities and choices, and the work is tangible—you build a lasting monument to your skill.

Although this book concentrates on bricklaying, it also includes information on the other trowel trades, such as becoming a cement mason, stonemason, tile layer, marble setter, and others.

CONTENTS

About the Author iii

Foreword v

Introduction vii

1. **What is a Mason?** 1
   What does a bricklayer do? You are considered skilled. Where does a bricklayer work? Masonry is more than bricklaying. Bricklaying leads to many other positions.

2. **History of Masonry** 5
   Masonry has a profound effect on the world. How long has masonry been used? Mortar was once just mud. Building well has always been important. What is a freemason? The masonry industry technology is growing.

3. **Types of Masonry Careers** 10
   What is a typical career progression? Masonry research keeps masons on their toes. Stonemasons. Cement masons. Mason tenders or helpers. Plasterers. Marble setters. Stone suppliers. Tile

layers and terrazzo workers. Brick sculptors. Union masonry instructors. The Peace Corps needs you. Chimney sweeping—a lucrative sideline or career.

4. **How Much Can I Earn?** 20

Pay is high for skilled workers. Cement masons. Tile layers and terrazzo workers. *Money* magazine rates construction pay high.

5. **Typical Workday for a Mason** 24

How does a building get built? What job conditions will I work in? Who is in charge? Mostly residential work. Daily decisions. Who will I be working with? Are minorities well accepted? What tools do I use? What do I wear? When do I retire? A cement mason's day. A stonemason's day.

6. **Advantages and Disadvantages** 36

Your skills will never be obsolete. Advantages of contracting. Masonry keeps you fit. Current image needs a lift. Safety needs to be a foremost concern.

7. **Is Masonry Right for You?** 43

Masons have a lot of pride. Masons have good math skills. Masons prefer working outdoors. Most masons live in or near a city. Becoming a contractor requires an aggressive personality and hard work. Masons must be fit and healthy.

8. **Future Growth of Industry** 46

Why get training after high school? There is a shortage of masons. Where are bricklaying jobs in this country? The union helps you find where the jobs are. Brick use is favored. Masonry unit manufacturing is consolidating.

Contents xi

9. **Union versus Nonunion** 55
   Do I go union or nonunion? Educating masons. Residential masonry is usually best for nonunion masons.

10. **Union Apprenticeships** 60
    What is an apprenticeship? Why have an apprenticeship? Who does an apprentice work for? Who pays for this apprenticeship training? Union bricklaying training is consistent across the nation. Where do I get apprenticeship information? How do I become an apprentice? What will I learn in an apprenticeship program? Other trowel trade apprenticeships. Tile laying and terrazzo apprenticeship. Cement masonry apprenticeship. Marble setting apprenticeship.

11. **Nonunion Training** 76
    Pre-apprenticeship program provides basic skills. Some local nonunion programs. Get an education while rebuilding America. Training for concrete masonry sales. How to become a certified masonry inspector. You can get college education in construction.

12. **Job Corps Training** 83
    What is the Job Corps? How do I enter the Job Corps? What will the Job Corps do for me? Where do I live while in the Job Corps? What do Job Corps classes teach me? What happens after I graduate from Job Corps?

13. **Earning Fame and Other Certifications** 92
    Contests for skill and design. Certification programs.

Appendix A: Glossary 96

**Appendix B: Bricklayer Union Bureaus of
Apprenticeship and Training** 103

Appendix C: State VICA Directors 109

Appendix D: Job Corps Offices 114

Appendix E: Industry Associations 126

Appendix F: Publications for More Information 144

Appendix G: Industry Trade Shows 146

CHAPTER ONE

WHAT IS A MASON?

WHAT DOES A BRICKLAYER DO?

Bricklayers, who are also called masons, put bricks and other building materials into place one at a time, with precision and skill. In 1988, there were 160,000 to 180,000 bricklayers in the United States, laying seven and one half billion bricks a year and four billion concrete blocks. Almost thirty percent of brick masons are self-employed, according to the Bureau of Labor Statistics.

Bricklayers build structures such as walls, floors, fireplaces, and buildings which range from basements to high-rises. Masonry walls can be structural or nonstructural and include all types of commercial, industrial, and residential structures which range from one to sixty stories high. Bricklayers build decorative chimneys, fireplaces, swimming pools, paving, arches, retaining walls, sun screens, fountains, murals, and playscapes. They work with brick, block (concrete, clay, and pumice), structural tiles, terra-cotta, and stone (natural or manufactured, both shaped and rough). These units are usually laid in a mortar or adhesive. This honorable craft provides a lot of satisfaction for people who want to work outdoors with their hands as well as their heads.

YOU ARE CONSIDERED SKILLED

It may surprise you to learn that two out of three construction workers are considered skilled workers. That is higher than almost any other industry, including manufacturing, where only one in five are skilled. Skilled means that you received specialized training in order to learn and master the techniques involved in performing your job. You are not just working by the sweat of your brow because you are also able to use your imagination to create and your logic to find the most cost-effective or easiest way to build a sound structure.

It was once said that "The man who has a trade has an estate." That means that if you have a skill that's needed—and masonry will never be out of date—you will always have the ability to clothe, feed, and provide shelter for yourself and your family. With today's high wages for skilled masons, you will have life's luxuries as well.

Your position will never be automated in your lifetime. You will always be guaranteed work because people will always need something built and there are many material advantages of brick.

Each brick you set in place does make a difference. Masons can always point to a place in the wall they put up and tell you where they didn't quite get the brick in place correctly. And it bothers them—because that wall will be standing there for a long time as a testament to both their skill and artistry and their occasional mistake.

The trade of masonry offers many challenges: laying the different materials into place so they last; reading blueprints and estimating job costs; and mixing mortar to meet specifications and weather conditions. Each job offers a new problem to solve or a new detail to build. Architects rely on a mason's knowledge to make the details work.

Masonry is humanizing because every brick is laid by hand. In his 1972 book, *Working,* Studs Terkel records the words of a mason named Carl Murray Bates:

"I can't imagine a job where you go home and maybe go by a year later and you don't know what you've done. My work, I can see what I did the first day I started. All my work is set right out there in the open and I can look at it as I go by. It's something I can see the rest of my life. Forty years ago, the first blocks I ever laid in my life, when I was seventeen years old. . . .

"Immortality as far as we're concerned. Nothin' in this world lasts forever, but did you know that stone—Bedford Limestone, they claim—deteriorates one-sixteenth of an inch every hundred years? And it's around four or five inches for a house. So that's gettin' awful close." (Laughs.)
—Copyright © 1972, 1974 by Studs Terkel. Reprinted by permission of Pantheon Books, a division of Random House, Inc.

WHERE DOES A BRICKLAYER WORK?

Masons usually work for a contracting firm. A contracting firm signs a contract that promises to build something for a certain price during a certain time frame. Contracting firms can work directly with an owner or indirectly through a larger contracting firm—usually called a general contractor—or through a developer.

Masonry contracting firms are typically small; the average company has only six employees, according to Bruce Voss of the International Masonry Institute. The construction industry is dominated by small and middle-sized contractors.

Nationwide, about 25,000 masonry contracting firms, including stone erectors, employ about 110,000 union masons and 200,000 nonunion masons. About 300,000 tenders and laborers are employed by these contractors to assist the mason. These contractors install about twenty-two billion dollars worth of masonry each year.

If you are a permanent mason on staff, you get to know the people you work with very well. If you free-lance between firms you'll meet a wider variety of co-workers.

Larger contracting firms called general contractors usually hire subcontractors, such as a masonry contracting firm, to do the brickwork and other specialty work. But some general contractors have their own masons on staff. In 1985, the National Association of Home Builders did a survey that found more than two-thirds of all general contractors and building firms use subcontractors.

MASONRY IS MORE THAN BRICKLAYING

Masonry is more than setting bricks in place. Masonry (never call it ma-son-ar-y—the word *masonry* has only three syllables) encompasses all the trowel trades: bricklaying, stone laying, cement masonry, tile and terrazzo work, paving with brick and precast units, and plastering and stuccoing. Masonry, stonework, and plastering are the second fastest growing small businesses according to the U.S. Small Business Administration.

BRICKLAYING LEADS TO MANY OTHER POSITIONS

People often begin a masonry career as an apprentice to a contractor and then become a journeyman. Some people stop there, content with the hands-on artistry of laying brick. But other positions are available; being a journeyman is not a dead-end job. If you like, you can work into the job of a foreman, a superintendent, an estimator, and in many cases you can own your own business as a contractor. You and your skills decide where you stop.

CHAPTER TWO

HISTORY OF MASONRY

MASONRY HAS A PROFOUND EFFECT ON THE WORLD

Did you know that seventy percent of buildings in the world are built of masonry? Stone and clay have always been a plentiful building material. Many great ancient structures were built of masonry. Examples include: The Great Wall of China, Rome's Colosseum, Peru's Machu Picchu, India's Taj Mahal, and Cambodia's Angkor Wat.

Masonry, such as brick and stone, is used because of its beauty, versatility, insulation, and because of its resistance to fire, wind, earthquakes, and sound. Its life cycle costs are lower than almost any other comparable building material. Over time, it is less expensive to own a masonry building than one made of other materials, such as wood or steel. Insurance costs are lower, and repairs and upkeep are minimal and infrequent.

Jack Hartray of Nagle Hartray Ltd., Chicago, summed up the historical importance of masonry at the Fifth North American Masonry Conference (NAMC) in June 1990:

"For most of western civilization, masonry and architecture have been identical. Masonry is architecture. . . . The quality that masonry has for me that I don't sense in other materials . . . is the ability to transmit an enormous load of cultural information . . . [Many modern buildings, in comparison,] have become sterile forms with very little stored information.

"We probably determined our value by what they've left in the way of masonry construction. . . . The sheer weight of masonry gives it a kind of seriousness—you don't build a cathedral unless you plan on leaving it to succeeding generations. . . . The marvelous plasticity of brick; it's a very expressive material. . . . American cities are largely masonry cities with an eruption of metal-framed structures at the center. . . . When you build you're sending a valentine to your grandchildren."

HOW LONG HAS MASONRY BEEN USED?

Bricks have been used for five thousand years. Brick is probably only second to wood in its long history of use. The art of brickmaking dates from 3800 B.C., as given on a brick tablet during the time of Sargon Akkad, founder of the Chaldean Empire in the Middle East. On the banks of the Tigris and Euphrates Rivers, sunbaked and irregularly cracked clay blocks were noticed and shaped by the people as a building material.

Around the same time as the building of the Tower of Babel, the Chaldeans learned how to burn brick in kilns, thus converting the clay into a hard substance. In the time of Nebuchadnezzar (604–652 B.C.), the Babylonians and Assyrians had acquired the art, not only of making hard burnt brick but also of enameling.

It is likely that the Chinese learned brickmaking from these early peoples. The Great Wall of China (210 B.C.) is partly constructed of brick, both burnt and unburnt.

The first brick buildings in America were erected on Manhattan Island in 1633 by a governor of the Dutch West Indies Company. The brick was made in Holland where the industry had long reached an excellent standard. By 1650, Americans started making their own brick.

Contemporary brickmaking is invariably automated. Getting handmade bricks is difficult. The few older masons who still practice the craft only produce limited quantities. However, a mason's skills are critical in developing nations, because making bricks by hand is less expensive than automation.

MORTAR WAS ONCE JUST MUD

Mortar is often called mud by many masons. That's probably because the earliest mortars were mud or clay. When the western territories of the United States first became populated with Europeans, many of them built mud houses which eventually washed away. When they built log cabins they put mud in the crevices to keep out the wind and rain.

Mortar is not made of mud. It is a combination of cement, sand, and water. If you were to add aggregate (stones or rocks) to it, it would become concrete.

Concrete is not cement. Cement is an extremely fine powder that undergoes a chemical reaction with water and when it hardens, it resembles the hardness of stone. Because cement is expensive and sand is a cheap filler, sand is added to the mix of cement and water. This also makes the mortar material more workable for laying bricks.

Many times you'll hear or read about a cement truck or a cement road. In reality, there is no such thing. It is a ready-mixed concrete truck and a concrete road. Remember, cement is just the powder.

Mortar is supplied in one of three ways: conventional, ready-mixed mortar delivered by truck (uncommon), and on-site continuous mixers. The conventional method requires a mason tender or other laborer to shovel the sand into the mixer with cement and water to make mortar. Other ingredients such as a powdered coloring agent or admixture for bonding strength may be added. If the crew is small, the masons may mix their own mortar.

BUILDING WELL HAS ALWAYS BEEN IMPORTANT

Today, we use building codes to regulate the way structures are put up. Depending on the part of the United States you are building in, you'd follow one of these three codes: the Uniform Building Code, the Standard Building Code, or the Basic Building Code.

Building codes, however, are not a modern invention. The earliest recorded building codes come from Babylonian King Hammurabi in the eighteenth century B.C.: "If a builder has constructed a house for a man, and his work is not strong, and if the house he had built falls in and kills the householder, that builder shall be slain."

In Polynesia, law required the builder to entomb a live slave under each corner post. This was meant to guarantee that the structure would be supported properly forever.

WHAT IS A FREEMASON?

Throughout much of history, bricklayers were called freemasons. They were the traveling skilled workers who built with brick. They were "free" to move about and were not tenured to anyone. They were respected for their craft and eventually a mystique grew around them.

Today, there is an association of "masons"—a secret society of sorts—many of whom have never put a brick in place and it is unrelated to the craft of masonry.

THE MASONRY INDUSTRY TECHNOLOGY IS GROWING

Construction from even twenty years ago differs from today's construction industry because improvements in material handling, packaging, and material ingredients have been made. New buildings also have better insulation and drainage, requiring more detailed and technical skills by the masons.

Although the basic skill of laying one brick atop another will continue to be the hallmark of building with masonry, the industry is still growing in technology and engineering. New insulations, mortar admixtures, and ways of constructing a structurally sound building are always in the process of development. Most of this work is done by the various industry associations.

CHAPTER THREE
TYPES OF MASONRY CAREERS

The U.S. Department of Commerce estimates that there are about 200,000 people working as trowel tradesmen. Add to that about 20,000 masonry contractors and many manufacturer and distributor representatives who entered the field as bricklayers.

Masonry is more than setting bricks into place. It encompasses a whole range of trowel trades and positions of leadership, administration, and even more specialized craftsmanship.

People who enter the work force today are likely to change jobs every four to seven years. The kinds of employment opportunities that we have are more flexible today. Opportunities in masonry offer flexibility as well.

Masons may advance to a position of foreman, superintendent, or estimator. A hardworking, aggressive person can easily become an independent contractor and employ other masons if needed.

WHAT IS A TYPICAL CAREER PROGRESSION?

The following list does not indicate a progression of better positions nor does it indicate progressively higher pay. It is just one possible progression if you decided to vary your masonry career. A brick masonry career can start with a job as a mason tender or a mason apprentice and then move on to any one of the following:

- journeyman brick or stonemason (Many people enjoy this position so much that they stay here until retirement.)
- mason job site supervisor or foreman
- quality control technician
- masonry superintendent
- estimator
- project manager
- masonry subcontractor or general contractor
- masonry teacher/instructor
- masonry product sales
- brick sculptor
- masonry inspector

You can make a lot of money in masonry or you can lose a lot of money depending on the local needs, your ability, and your desire to work hard. If you are a contractor, your business skills greatly affect your profitability.

Here is a list describing some of the other positions available in contracting:

- The foreman is a skilled journeyman who supervises the activities of all the bricklayers at the job site. He or she schedules workers, plans the work to be done that day, ensures productivity, and reports to the superintendent.
- The superintendent can come in on either the job or general level depending on the size of the contractor or the size of the job. The superintendent carries out directions of the project manager and contractors.
- The project manager determines schedules and procedures for a job. If more than one trade is involved, he or she coordinates the timing of their efforts. Other duties include: reviewing

and scheduling material deliveries, establishing priorities, and obtaining clearances.
- The estimator obtains information about a job from the plans and specifications concerning materials, equipment, and type and hours of labor needed. Using this information with input from the contractor and other staff, the estimator can competitively price a job for bidding. You also determine the most economical way and source to buy materials and store them. In a very large contracting firm, a purchasing agent would do the actual buying.

MASONRY RESEARCH KEEPS MASONS ON THEIR TOES

J. Gregg Borchelt, director of engineering and research at the Brick Institute of America, says there is a lot of research going on in the masonry industry. At his association he is looking into new products such as "interlocking brick pavers that can take loads without a concrete base; brick units whose properties vary through the wall; products that change properties on demand (for example, a product that lets heat in when you want it, but keeps heat out when you don't); an electronic barrier inside the wall that can repel water; new shapes that permit running vertical and horizontal fiber optics, gas lines, and communication lines; additives to reduce efflorescence problems; and new bonding agents—perhaps an adhesive tape that's applied to brick and cures on-site."

Borchelt also sees increased competition from customized concrete masonry (because he is in the brick industry); continued competition from vinyl and aluminum siding; serious competition from gypsum board, once a durable, exterior-finish board is developed; tighter specifications on brick color, chips, cracks, and blends; quicker test procedures that give more reliable results; conversion to metric mandated by government; more emphasis in model building codes on alarms and detectors than on materials; code adoption of strength design standards that allow more economical masonry design.

Masonry techniques and design do change and testing is being done by various associations for the following: better ways to construct floor diaphragms; new ways to make grout test specimens; and testing two-story shear walls with openings to find lateral load capacity and failure response.

STONEMASONS

Stonemasons spend six thousand hours learning to use tools to set stones, polish stones, drill in masonry, chip stones with a chisel, and set substitute material. As a stonemason, you are considered an artisan.

Stone erectors are considered a specialty contractor. Before 1975 most stone was set in mortar. Now about ninety percent of all stone veneer is set without mortar or grout. They are connected with mechanical anchors.

CEMENT MASONS

Cement masons work with concrete, wood, and metal to create floors, sidewalks, roads, walls, basements, patios, and driveways. On the job a cement mason will:

- mix and place concrete
- determine quantity of materials needed and placement of joints
- ensure concrete is level or that pitch is correct
- use power equipment such as troweling machines, drills, and jackhammers
- create different textures in the finish

- apply various finishes to concrete such as color or acid etching

Cement finishers are in high demand because although it is a highly skilled art if done correctly, it is also extremely difficult work. You are often on your knees to apply hand finishes. Many long-handled tools help to make the cement finisher's job easier, though.

You must be physically fit and willing to work hard. Some cement masons retire in their late forties or early fifties because the work is hard on your back and uncomfortable if you have arthritis. In Chicago's union, the retirement age is sixty-two; if you work fifteen consecutive years and reach age sixty-two you can retire with full pension benefits.

After working as a journeyman, other career paths for a cement mason (finisher) include foreman, superintendent, quality control technician, and ready-mixed concrete dispatcher who schedules and coordinates the concrete trucks' arrival to job sites. You can become a contractor or go into sales or estimating too.

MASON TENDERS OR HELPERS

To become a mason tender or helper takes only two thousand five hundred to three thousand hours of apprenticeship, in not less than one year. This position trains you to mix mortar, carry mortar and brick to the mason, strike off completed joints, and caulk. The helper also wets and cuts brick as needed and cleans up after the masonry work is completed.

This position is an excellent entry level position if you are undecided about which building trade to enter. It does not pay comparably to a mason's job, but it does not require as much skill either.

PLASTERERS

Plasterers apply plaster or stucco to inside walls and ceilings or outside walls. Plaster contains gypsum, lime, portland cement, and water. In interior work, these skilled workers apply three coats: a scratch coat, brown coat, and a finish coat. Plastering lets you express yourself with textures and ornamental plaster castings such as medallions.

To become a journeyman plasterer, a person must work three to four years at the craft in a training or apprentice program. An apprentice supplements this practical experience with 144 hours of classroom instruction in subjects related to the trade for each year of apprenticeship training.

In a typical four year program, apprentices learn how to use and care for tools, how and when to use various materials, how to do casting, molding, and some ornamental work, how to apply plaster, how to use a plastering machine, how to set levels, and how to cost a job and time it.

On the job, the plasterer's apprentice assists the journeyman in all straightening, filling in, and finishing operations after the journeyman has completed the critical elements of the work. At first, the apprentice performs tasks usually done by a tender or laborer such as mixing plaster, building scaffolding, handling materials, and cleaning up.

As a journeyman plasterer, you will:

- mix plastering materials
- construct cornice molds
- apply veneer plaster, stucco, and exterior insulating finishes to create straight and level walls and ceilings
- create different textures

After becoming a journeyman plasterer, other career options are also available. Like the other trowel trades, you can move up in the ranks of your contracting firm. You can also go into commercial art and theatre to build castings for various productions. Even Walt Disney Studios hires plasterers for its art needs.

MARBLE SETTERS

Marble setters usually don't work for a contractor. They usually work for a firm that manufactures or distributes marble. They work indoors in residential and commercial settings. The crews usually only include a journeyman or two, and a helper who is learning the trade, possibly as an apprentice. In 1990, marble setters made about $400–$700 a week.

STONE SUPPLIERS

As a stone supplier, you help masonry contractors by establishing credit with foreign banks, reviewing and approving samples, making material quantity take offs, ensuring enough inventory of the stone, giving packing and crating instructions, and providing shipping documentation. You are the liaison between a foreign stone producer and the contractor.

TILE LAYERS AND TERRAZZO WORKERS

Tile layers and terrazzo workers cover walls, floors, ceilings, and other surfaces with these materials. Terrazzo is concrete that has marble chippings as aggregate. Terrazzo workers put prefinished materials into place by drilling holes for anchors and using adhesives to fasten them. Tile layers cover walls, floors, and ceilings with tiles.

BRICK SCULPTORS

Brick sculptors are artists who sculpt brick to create designs on walls. Some brick sculptors start as masons, others start as artists working in another medium. This art form has had a resurgence of popularity in the last fifteen years according to J. Gregg Borchelt, the director of engineering and research at the Brick Institute of America.

To make bas-relief brick sculptures, thick bricks are carved green, then taken apart and fired before being installed on the job site. In the ancient days, brick sculpture depicted figures of cultural and religious significance on monumental Babylonian and Mesopotamian structures in about 600 B.C. Now they depict anything from the abstract to animal life and corporate logos.

UNION MASONRY INSTRUCTORS

Instructor training is crucial and instructor certification is key to the Trowel Trades Training System. To become a Certified Instructor of Journeymen and Apprentices in the Trowel Trades, a candidate must complete 160 hours of course work and demonstrate proficiency in technical tasks and teaching skills.

For the course work, instructors must attend a one-week training seminar each summer for four years. Courses at this annual instructor program range from education basics, such as principles of adult learning, to advanced masonry topics, such as solar masonry design and masonry insulation.

Since 1986, nearly six thousand people have taken part in this new training program. The Brick Institute of America has also launched an annual award for bricklaying instructors. Any full-time bricklaying instructor in a high school, area vocational school, or community college is eligible.

THE PEACE CORPS NEEDS YOU

If you have a desire to become a vocational teacher or worker in an underdeveloped part of the world, apply with the Peace Corps. This governmental organization desperately needs people with building skills.

Life in the Peace Corps can be satisfying. You are helping nations build an infrastructure that will continue beyond your initial helping hand. But the Peace Corps can be frustrating as well. Many times the work that needs to be done in these nations is overwhelming. Your accomplishment might seem small in comparison with the work that is needed but to the people you are helping, it is significant and worthwhile.

You are paid a low wage in the Peace Corps and usually live in substandard housing conditions compared to life in the United States. Your experience with the Peace Corps, though, will enhance your views of life.

CHIMNEY SWEEPING—A LUCRATIVE SIDELINE OR CAREER

The homeowners' love affair with fireplaces is strong, creating a need for the flues to be cleaned. Cleaning a chimney and fireplace helps keep a fire in the fireplace instead of in the house. Ignitable substances, such as creosote, build up inside a fireplace and chimney after many uses. The chimney sweep cleans the inside surfaces to keep the fireplace safe. They also repair chimneys. There are about seven thousand five hundred chimney sweeps in the United States.

Chimney sweeping is not seasonal, though it is perceived as such by the public. After three to five years in the business, a chimney sweep business owner should have a marketing plan in place that guarantees him or her full-time work for the entire year.

No state requires any training to become a chimney sweep, nor is there a union for the trade. Many chimney sweeps are the owners of their business with their spouses as business manager. Although sweeps usually work on their own, the business likely has employed one other sweep and up to twenty sweeps in a larger business is possible.

Training is not needed to become a chimney sweep, but it is recommended by the National Chimney Sweep Guild (an association, not a union). Through its nonprofit, tax-exempt educational arm, the Chimney Safety Institute of America, training and certification is available. It is a home-study program that involves attending weekend seminars and has a six-day program available that is usually held at various college campuses.

To learn how to enter this industry, either as a career or lucrative sideline, contact: National Chimney Sweep Guild, P.O. Box 429, Olney, MD 20832 (301-774-5600). They will send you brochures and other information that will tell what a sweep does, what tools and equipment are needed, and where to get training.

CHAPTER FOUR

HOW MUCH CAN I EARN?

PAY IS HIGH FOR SKILLED WORKERS

Wages for bricklayers are near the top of those in the building trades and rival wages for skilled workers in other industries. If you are a union bricklayer you usually earn more than a nonunion bricklayer.

Bricklayers are paid by the hour, rather than by a weekly, monthly, or annual salary. But their benefits rival salaried workers', especially if the bricklayer belongs to the union.

In May 1991, here's how the trowel trades (union rates) compared with one another:

- Bricklayer: $21.13 per hour, with $4.16 per hour in benefits (For comparison: in February 1987, bricklayers earned $13.25 an hour, or $455.20 a week. In May 1993, a bricklayer in the Local #2 or #13 Unions in Southern California earned $26.37 per hour.)
- Cement mason: $20.75 per hour, with $5.42 per hour in benefits (In August 1992, a cement mason in Chicago earned $22.00 per hour.)
- Plasterer: $17.25 per hour, with $3.10 per hour in benefits

- Marble setter: $21.73 per hour, with $4.35 per hour in benefits
- Terrazzo worker: $20.30 per hour, with $3.80 per hour in benefits
- Tile layer: $20.00 per hour, with $4.25 per hour in benefits
- Tuckpointer: $21.10 per hour, with $4.36 per hour in benefits

Benefits refers to the contribution made to welfare (insurance), pension, apprenticeship training, promotion, and other funds. In most cases, if you do not work, you are not paid. That means you do not get paid holidays, vacations, or personal days.

Overtime is usually paid at double time and you are not forced to do piece work—all work is paid by the hour. Foreman wages are seventy-five cents to two dollars per hour over wage scale, depending on crew size.

All journeymen in your local union earn the same rate. If you are a journeyman, you do not have to worry that someone is making more for doing the same work that you are.

In addition, the masonry contractor employing you pays for your local and international pension; your medical, dental, and vision insurance for you and your family; and a portion is given to the union to help cover expenses such as apprenticeship programs.

The bricklayer union initiation fee is $400.00 ($200.00 down and $200.00 within 60 days). Union dues in 1992 were $32.95 per month plus a working dues of $0.35 per hour.

Bricklaying apprentices begin work at forty percent to fifty percent of the journeyman's wages with a pay raise every six months until they reach ninety-five percent of the full pay in the last six-month period of the three-year apprenticeship.

CEMENT MASONS

Cement mason apprentices begin work at a percentage of the journeyman rate, but that percentage varies. In Lake County,

Illinois, apprentices start at forty percent of the journeyman's wage, but in Cook County (Chicago), they start at seventy percent of the journeyman's wage.

Not all cement mason unions offer the same benefits. None of them provide paid days for holidays, vacations, or personal days. If you don't work, you don't get paid. The average cement mason works about one thousand five hundred hours a year because of the seasonality of the work. Normally, you can't place concrete in extreme weather conditions. (A nine to five type of job takes two thousand hours a year.) For bricklayers it varies—some masons work throughout the entire year—others are affected by weather constraints, lack of building in a part of the country, or other factors.

TILE LAYERS AND TERRAZZO WORKERS

In 1990, most terrazzo workers, tile layers, and tile setters earned $400–$700 per week. Their work may continue throughout the entire year because it is primarily indoors. Almost fifty percent of all tile layers are self-employed according to the Bureau of Labor Statistics.

In general the apprentice receives about fifty percent to sixty percent of the journeyman's pay rate at the beginning of training. During the three years of apprenticeship for each of these specialties, the rate can gradually increase to ninety-five percent of the journeyman's rate.

MONEY MAGAZINE RATES CONSTRUCTION PAY HIGH

In February 1992 *Money* magazine ranked one hundred jobs for earnings, security, prestige, and satisfaction, and gave an outlook for 1992 and a growth figure over a period of fourteen years. For a construction worker, the magazine ranked the job at ninety-six out of one hundred as desirable on the list. Ratings of other positions included: secretary at forty-three, journalist at eighty-three, and butcher at ninety-seven. Construction work was probably rated poorly only because of the public's general perception of all construction workers lumped together. A skilled worker—in bricklaying, carpentry, or pipe fitting—would have a higher ranking.

The magazine rated the ninetieth percentile annual pay for a construction worker at $36,364. Security, prestige, and satisfaction were rated as poor. Again, this is because of the cyclical nature of many construction positions and because some construction jobs don't require much skill. The 1992 outlook was rated good with a fifteen percent growth over a period of fourteen years.

Unfortunately the magazine did not specify masons in their own category. Their security, prestige, and satisfaction would rate higher because bricklaying is a well-respected and highly skilled trade that always has work in some part of the country.

CHAPTER FIVE

TYPICAL WORKDAY FOR A MASON

The typical workday begins early and masons are usually done by three o'clock or four o'clock in the afternoon, unless overtime is needed. Coffee breaks are important to masons so they can rest for a few minutes. These breaks usually occur when building supplies are being replenished or some other activity precludes them working for a short period. Masons bring their own refreshments and food to the job site.

There is no bell or whistle to let masons know when it is time for a break or lunchtime. The foreman usually just calls out when it is time to come off the wall. If you are working by yourself or with a co-worker, you generally decide yourself when to take a break.

HOW DOES A BUILDING GET BUILT?

To put up a wall, the foreman and the masons first look over the blueprints and specifications together. Most commercial or industrial brick buildings consist of concrete blocks as a backup wall and bricks as a facing wall. Homes usually have a wood frame backup with a brick full or partial veneer. This meeting is similar to a football coach getting his players ready before the big game. It helps get everyone off on the same foot.

The next step starts the actual building. Oftentimes, the foreman builds the corners at each end of the wall. This gives the masons a starting point for getting a wall up.

Masons use a plumb line and a level to keep the wall square and straight. A line stretched between these corners also serves as a guide for each row (course) of bricks. The masons match the top corner of the brick with the level of the string to keep the wall straight.

Each course of brick are laid in a layer or bed of mortar. After they spread a bed of mortar, the masons put mortar on one end of a brick and place it on the bed. They press the brick in place, lining it up with the string guideline, and then slice off the excess mortar with a trowel.

Some procedures—such as laying a bed of mortar (called throwing a mortar line), for example—may seem to be tricky. A mortar line is the ridge of mortar laid on top of a course of brick for the next course of brick to be set on. This technique involves twisting a trowel so that mortar spills off to form an even ridge. Doing this quickly and correctly takes practice.

Standard brick is nominally eight inches long and four inches wide. Bricks and blocks are usually delivered to the scaffold by forklift on wooden pallets. A steel brick carrier called a hod can be used to pick up ten bricks at a time.

Masons also cut bricks with a trowel, brick hammer, brick set or masonry saw to fit the materials around windows, doors, pipes, and other openings. If required in the design, masons build brick or stone arches over windows and doors.

Bricklayers finish the joints between bricks with a jointer, a tool that compresses and finishes mortar joints. The finish of the joint is critical to giving the wall a neat appearance and making it watertight.

To clean the finished work, they use a brush over the wall to knock off any mortar that was not compressed with the jointer. After the mortar has set they may use water or an acid base solution to further clean the wall. Masons also caulk the spaces around doors, windows, and other openings.

Some bricklayers just lay brick upon brick on a stright wall all day long. These masons are called "line burners." The mason tenders ensure that they have a constant supply of mortar and brick and the foreman ensures that the scaffold is always at the best level. These masons are efficient and fast. Their productivity and skill is high, but their creativity is low.

WHAT JOB CONDITIONS WILL I WORK IN?

Masons mostly work outdoors in all kinds of weather, maybe high on a scaffold or in tight damp quarters of a basement. In cold weather, protective sheeting is often used to enclose the working area to keep masons working throughout the year.

Weather often determines if you work or not that day. If winds are over twenty-five mph, most contractors take everyone off the job because dust gets in your eyes and high-rise work may be unsafe.

Most work will be in or near cities, but masons work all over the country. Many times stonemasonry is required in rural areas.

Job hazards include falling or having something such as a tool or brick fall on you. Wearing hard hats and steel-toed shoes helps to protect masons. If work is going on above you at the job site, nets or wood planks placed above you are often used to catch falling objects.

Overtime work is not uncommon in masonry because of the shortage of qualified masons in many parts of the country. Masons like working overtime because they are usually paid twice their regular rate. Many masons do odd jobs on weekends.

WHO IS IN CHARGE?

There is a chain of command on the job site. It generally follows this order:

- general contractor
- specialty contractor
- project manager
- general superintendent
- job superintendent
- foreman
- journeyman
- mason tenders
- laborers

This chain of command may not have each of the rungs in place on every job site. For example, if you are putting up the brick veneer on the front of someone's home, it might be only you and a fellow worker (possibly the contractor) on the job site. Only very large masonry jobs would have a project manager, general superintendent, and a job superintendent.

The journeyman mason is not "in charge" of the tenders and laborers, but he or she can request work from them, such as bringing more materials to the wall or cleaning the wall.

MOSTLY RESIDENTIAL WORK

Typically masons do more than build straight walls. "About sixty-six percent to seventy-five percent of masonry work is done in residential construction, mostly by nonunion bricklayers," says Bruce Voss of the International Masonry Institute. That means that the crew is smaller and does many different job site duties. This variety of work includes:

- Tuckpointing
- Cleaning brick and stone: water, chemical or abrasive methods; using hand-scrubbing, spraying, pressure washing, steam, poultices
- Preparing corner leads
- Cutting brick/block on a saw
- Building a chimney and fireplace

- Adding brick design accents
- Making the mortar

On a commercial or industrial building, the work often requires adding polystyrene insulation and reinforcement to the wall. The reinforcement might be wire to hold two wythes of masonry together—such as a concrete block wall and a brick wall—or iron rods, called rebar, in the holes of the concrete block. Another specialized technique is grouting the holes in concrete blocks (by hand or by a pump or hose to hoseman on the wall). When rebar is used in the holes of concrete block, they are grouted into place.

DAILY DECISIONS

It is not unlikely for a foreman or architect to ask your opinion on the easiest or best way to build a detail on the wall. This is yet another area where your experience and skills are critical to ensuring a building's success.

Daily decision making often depends on the weather. If it is very hot, you'll have to be concerned with:

- Keeping the mortar moist (adding more water is called retempering)
- Not spreading the mortar bed more than four feet ahead of the masonry
- Applying a light fog spray during first twenty-four hours or using damp burlap over the way
- Keeping materials cool (shade the sand)

Cold weather, too, affects the way you work. At below forty degrees Fahrenheit precautions must be taken such as: keeping materials off wet ground; preventing mortars from freezing and cracking; and protecting your walls with tarps.

The Portland Cement Association found that workers stopped working on unprotected masonry when the air temperature was seventeen degrees Fahrenheit, but they could still work if on protected scaffolds until wind chill fell to minus two degrees Fahrenheit to minus seventeen degrees Fahrenheit.

Residential buildings may appear easy to build, but many times, brick homes have brick fireplaces and chimneys. These are not so easy to build. Many code requirements must be followed to build a masonry fireplace. However, many fireplaces today are not full-masonry fireplaces and require little skill to install. For a full-masonry fireplace, the mason must consider:

- Clearance from combustibles (such as supporting stud wall)
- Enough wall thickness so heat is contained
- Liner joints so smoke flows correctly
- Properly supported inner and outer hearths
- Proportions of the smoke chamber to help flow of gases
- Clearance between flue and chimney to avoid cracking
- Making the chimney tall enough to avoid wind-induced downdrafts

Each job has its own quirks and requires the masons to think out how to solve any problem that arises.

WHO WILL I BE WORKING WITH?

The masonry industry is dominated by contractors which are primarily small and medium in size. Crew sizes vary, but having five or so masons on a wall is not uncommon. Though usually part of a crew, each mason has his or her own duties or area of the wall that is being built. Masons seldom work alone unless they are doing small residential jobs.

Some masons always work for the same contractor. Others go where the work is and switch companies frequently. Others have a main contractor they work for, but when times are slow, they go where the work is. Self-employment is not uncommon for masons, especially those willing to work weekends.

ARE MINORITIES WELL ACCEPTED?

Minority groups such as African-Americans, Hispanics, and people from the Middle East are becoming more and more common on the job site. Those groups are forming a bigger and bigger percentage of the ethnic mix of this country so it is only logical that they become a bigger part of this skilled profession.

Whoever pulls their own weight on the job site is well-respected, regardless of color or creed. All apprenticeship programs actively recruit minorities. In addition, contractors cannot afford to discriminate—bricklaying is such a valued skill and so in demand that every skilled worker is needed.

Minorities are entering the masonry industry at a rate almost consistent with their percentage representation in this country. On the other hand, because masonry work has been done mostly by men throughout history, women are still not a common sight on the job site, but this is slowly changing. Many times, however, the only woman on the job site is the architect.

More and more women are discovering that construction offers greater financial rewards and more satisfaction than roles traditionally available to them. As economics require more and more women to work, this number will surely grow in the coming decade. For more information about career opportunities in construction for women, write or call the National Association of Women in Construction, 327 South Adams Street, Fort Worth, TX 76104 (817-877-5551).

Many men are finding it to be a refreshing change of pace to have female co-workers. It gives each sex a healthy appreciation of the other. Some women find that since they do not have to dress attractively as part of their jobs, men regard them as co-workers rather than objects. Most women in the construction trades are finding a warm reception, especially once they have proved they can pull their weight and be part of the team.

WHAT TOOLS DO I USE?

To do good work, a mason needs first class tools. A knowledgeable foreman can judge a new bricklayer by observing the condition, quality, and brand of the tools he or she carries. Compared with some other tradesmen, such as carpenters, masons need few tools.

As a rule, all the tools a mason needs can be carried in a canvas tool bag. It is a common sight to see masons climbing the ladder to the scaffold with a canvas bag supported over their shoulder by their four-foot level.

A basic tool set of about twenty-eight pieces will cost between four hundred dollars and five hundred dollars, including the bag. These tools usually include a four-foot level, a two-foot level, chalk, nylon line, trowel, pointing and bricklayer's trowels, folding measuring tape, brush, chisels, slickers, and various jointers and hammers. The brick trowel and four-foot level are the most used tools.

Masons most commonly use brassbound, four-foot mahogany levels to ensure their work is level and plumb. Aluminum levels do not always stay as true to level in rough job site situations. Some masons' levels last eight to ten years, but more often than not the levels only last a year because of rough treatment. A high quality level costs between fifty dollars and sixty-five dollars.

Other mason tools include: other types of trowels and hammers, rubber or wooden mallets, wheelbarrows, shovels, plumb bobs, and drills. Their machines include: hoists, pincers, follers, tampers, and polishing and grinding machines. Electric mortar mixers, power lifts, saws, and splitters make their work easier. The contractor you are working for will most often supply these power tools.

Masons take much pride in their tools. They keep their level oiled so that mortar does not stick to it and the glass vials stay clear. They wash their trowels and other tools after using them, again to keep them free of hardened mortar.

Masons take their equipment with them to the job site each day. They don't usually leave it on the job site. The care a mason takes with a level usually indicates the level of care taken on a job. If the mason leaves a level in the back of a car or truck, lying flat and exposed to the sun, it's likely the work that mason does is not of a high quality. If it is hung vertical on a nail, that is the mark of a conscientious worker.

WHAT DO I WEAR?

Masons wear hard hats when they have work going on overhead, which is most of the time. It is sometimes difficult for new masons to adjust to wearing the hard hat, but experienced masons know its value and keep it firmly in place. Hard hats are made of strong, reinforced plastic and they have a liner on the inside that holds onto your head. When you lean forward, the hat should be secure enough not to slip off.

Usually masons wear jeans or heavy cotton pants and long sleeved shirts. This protects their skin from abrasions when they may accidentally walk too close to a supply of bricks, over a pile of reinforcing wire, or other material on the job site.

Hard-toed shoes are essential for safety. OSHA—a safety regulating agency of the federal government—requires heavy-duty footwear because of the many hazards on a construction site. If you arrive to work in sneakers, you will likely be sent home without pay.

When it is hot some masons wear short sleeves or tank tops, but it isn't advised by OSHA. Some masons use gloves with open fingers to help protect their hands. Masons' hands often are strong and thickly calloused because of years of handling bricks and blocks one at a time over and over again.

When cutting bricks on a saw, masons wear plastic safety goggles and heavy gloves that cover much of their lower arm. Masons rarely wear hearing protection, but some job site activities are so noisy, ear protection would be made available from the contractor.

If working on a high-rise, occasionally masons must work on a tether with a body harness to protect them in case they fall. This is not often the case, though, because most mason scaffolds have a barrier rail to protect you from falling.

Using rapelling equipment (harness and rope) like a rock climber is a skill that can be learned. It is helpful for inspecting high-rise walls and doing repair work in areas of a building that are difficult to reach. It is not a common method of gaining access to a building, but it is effective for some types of work.

WHEN DO I RETIRE?

There is no set retirement age for masons. Many retire in their fifties or early sixties though. Masonry is hard work, albeit rewarding work, but oftentimes older masons choose to take a less strenuous position in masonry, such as superintendent.

The local union will have a set retirement age, usually around age sixty-two to sixty-five. When you retire at this age with the correct number of years of service, you receive full pension benefits. You do not have to be a bricklayer or a cement mason all those years leading up to retirement age. There are many less strenuous positions you can hold while working within the union.

A CEMENT MASON'S DAY

Most cement masons work on small residential jobs such as driveways, basement foundations, sidewalks, or patios. However, they also can be found placing the concrete into forms for bridges, high-rise buildings, and sports stadiums.

To place most residential concrete, the cement mason works with a small crew of masons usually consisting of two or three people. The masons put the wooden braces for the formwork into place with sledgehammers and the braces are also nailed into place. They ensure that the grading material for the concrete to be laid on is correct and to the necessary grade. They ensure the correct amount and specifications of ready-mixed concrete is ordered and that it arrives when they need it.

As the material comes down the chute from the truck, the cement masons guide it into place with shovels. They are careful when walking through the concrete with their high boots not to get concrete on their skin too much or on their clothing. Concrete is a caustic material and can produce a severe chemical burn if you leave it on your skin for too long.

When the concrete fills the formwork, the cement masons level it off with a screed, which is a tool that runs along the top of the form to knock off extra concrete and produce a roughly flat surface on the concrete. Then comes the finishing process. Finishing concrete is truly an art. Knowing when to finish with what tools takes a practiced eye and hand. If you finish the concrete too soon, too much water may be trapped below the surface of the concrete and will eventually result in a failed finish as the water tries to escape. If you finish the concrete too late, it might have hardened up too much and you will not be able to work it into place.

Cement masons use floats and trowels to finish the concrete. Finishing brings the fines in the concrete mix, such as sand, to the surface to create an attractive finish and one that is sealed against the elements. After the last trowel pass, the concrete must be protected from too much moisture escaping (it must be cured). Curing usually requires a wet burlap or straw covering, although a curing compound which can be sprayed on the concrete can be as effective. Some concrete is not cured because the weather conditions do not merit it, but that is not usually the case when placing high quality concrete.

On commercial or heavy construction job sites, such as bridges, the concrete construction is more detailed. Sometimes reinforcing steel or insulation is needed. Forms to contain the ready-mixed concrete are often prefabricated from a manufacturer.

These engineered projects take a lot of math skills beyond just ordering the correct amount of concrete. You are often working with an engineer or architect to ensure the strength needs of the structure will be met and that other trades' work, such as ductwork, will fit after you have put the concrete in place.

A STONEMASON'S DAY

Stonemasons build walls, buildings, and floors from stone. They work with natural cut stone such as marble, granite, limestone, or sandstone. They also work with artificial stone made of concrete or other masonry materials.

Sometimes the stones are numbered and the stonemasons use a blueprint to guide their placement. With large stones, power lift operators hoist them. Like bricklayers, stonemasons set the first layer in a shallow bed of mortar.

If you were a marble setter, you would have to be very meticulous in your work. Marble is a fragile, easily scratched material and usually must be handled by hand to avoid chipping. Since a one-inch marble panel weighs about seventeen pounds per square foot, often two or three marble setters work together to lift panels.

Marble setters mostly work indoors, on commercial sites as well as residences. In residential work, the journeyman—or mechanic, as some are called—usually works with a helper. Marble setters more often are employed through a marble manufacturing firm rather than a contractor.

CHAPTER SIX

ADVANTAGES AND DISADVANTAGES

YOUR SKILLS WILL NEVER BE OBSOLETE

Look around you—brick is everywhere. As a bricklayer, or any masonry specialist, your skills will never be obsolete. There will always be a need for masons.

In a 1987 survey of home buyers, *Professional Builder* found that half of all consumers would prefer a home with a brick exterior. Interestingly, though, only 35.2 percent of builders offer brick homes. Brick fireplaces outranked bay windows, concrete patios, and wood decks as first choice options for homes. Brick is here to stay.

Load-bearing brick walls can combine the structure, the exterior envelope, the interior partitions, and the inside finish in one trade. Having that cost-effective ability makes masonry valuable to owners and builders. That means there will always be work for you on many different job sites.

ADVANTAGES OF CONTRACTING

To become a contractor you do not need a license or specialized education. To be a good contractor, though, you will need skills and training in your area of construction and good business skills.

Since there are no requirements for becoming a contractor, though, many have little more than a pickup truck and a wheelbarrow. But these contractors seldom last more than a year or two. It takes a lot of skill and perseverance to become a successful contractor.

Opportunities to advance and own your own business are typical of the construction industry. But being a contractor, whether you are small or large, specialized or general, public or private, requires construction knowledge, an aptitude for handling a business, and lots of long, hard work. The failure rate of contractors is high compared with other industries, but at least you have the opportunity to try.

MASONRY KEEPS YOU FIT

Being a mason is strenuous work. You stand, stoop, kneel, bend, and reach. You get regular breaks, though, to rest.

Masonry's great strength and durability make it desirable and sometimes intimidating to those who must lay it. Each brick weighs about five pounds. A concrete block can weigh up to fifty pounds. Concrete itself weighs about 150 pounds per cubic foot. Lifting these elements hour after hour is hard work, but masons have many devices and techniques which conserve labor. Wheelbarrows, mechanized vehicles such as forklifts, and scaffolds raised to the proper work level conserve a mason's energy.

CURRENT IMAGE NEEDS A LIFT

Masons once enjoyed an elevated position in society. Their craftsmanship was regarded highly and membership in the mason's guild was sought after. Throughout history, people who excelled at building were highly respected members of society. That has changed in the last few decades because the current fashion dictates that desk jobs in an office are better.

But opinions do change, and many masonry associations are promoting change in an effort to regain the respect that masonry once had. Using billboards, meeting with high school students, and even having commercials on MTV have communicated to many young people that masonry construction offers satisfaction and high wages.

Recruitment is hard because of the negative image bricklaying has with the public. This image includes a second class image of vocational education. Also, there is a low initial salary for apprentices and disappointment with work conditions, especially in colder climates.

"Fewer people each year are willing to enter the United States masonry industry. The construction industry in general and the masonry industry in particular have a poor public image. Our industry is perceived as a business run by bulldozer mentalities and staffed by semiskilled personnel with no glamour in their future," says C. DeWitt Brown, a masonry contractor in Texas.

Today the perception is that a desk job and business suit bring you respectability. Image is derived from attitude. The various associations in the masonry industry are out to change attitudes. The Brick Institute of America (BIA) and the National Concrete Masonry Association (NCMA) have produced a ten-minute videotape for high school students to promote a career in masonry. The title is "Bricklaying—Master the Craft and Make Your Mark" and costs thirty-five dollars from BIA or NCMA. These associations have also produced posters and booklets that show the pride and accomplishment that is gained by becoming a mason.

The masonry industry has been its own worst enemy in the past. Rather than shine a bright light over the skill and tout it publicly, the industry associations have ignored the value of good public relations. This image that society has thrust upon the masonry industry also affects the productivity of masons on a job site. Some masons do not feel the pride they should because they come to believe they have second class jobs.

James J. Adrian of Adrian and Associates, Peoria, Illinois, says that craftsmen need to feel part of a team to be productive. He says that workers actually are working only four hours in an eight-hour day. One third of the problem is because of poor management, poor scheduling, inadequate training, and no standards for productivity.

This is in the process of changing. As more and more people in the industry open their minds to change and productive, safe ways of working, morale is improving. Magazines, association newsletters, and trade shows are now coming into the forefront to educate contractors and masons and revive the pride that was once theirs. Once the masonry industry was silent with the government. Now the NCMA has a full-time lobbyist in place to heighten awareness of the masonry system.

Bruce Voss from International Masonry Institute (IMI) says:

> "A young person has many reasons to pursue a career as a mason, but year after year one of the most common ones is the pride and satisfaction that comes from building. When the job is done, you know you've accomplished something. You haven't spent all day connecting Part A57 to Part X93. Instead, with your hands and learned skills, you've built something that will endure. You've left your mark. What before was only a drawing on paper is now a lasting part of the community. You can step back and say 'I helped build it.' And years later, someone else will echo the words: 'My grandpa helped build that.' "

The distinctiveness and value of masonry depend on its craftsmanship, not just on the quality of masonry products. It is craft and product together that give masonry its place in the building market. Diminishing the craftsmanship aspects of masonry also diminishes its appeal. Built brick by brick, block by block, stone by stone, masonry is a personal material that people relate to on a human scale.

SAFETY NEEDS TO BE A FOREMOST CONCERN

Depending on the source you read, construction is the first, second, or third most dangerous industry to work in. It is always in the top three though, along with farming and mining.

Construction had the highest number of job related deaths from 1980 to 1984 with a total figure of 952. It also has the second highest death rate: 23.1 workers killed per 100,000. The mining industry has 30.1 per 100,000 workers. Nationally the average yearly number of job related deaths was about 7,000 and the average yearly death rate was 8.8 per 100,000 workers.

Work related accidents account for thirty percent of all accidents and cost thirty-five billion dollars a year and that is only the direct costs. Indirect costs come from workers who need to be replaced, code violation fines, accident investigation cost, and overtime to make up for loss. Business management studies reveal that indirect costs of accidents cost four to ten times as much as direct costs and are not recoverable by insurance.

Workers' compensation rates for construction are higher than other industries. A contractor can pay up to fifty dollars for every one hundred dollars of payroll expense. That is because construction can be dangerous.

Special trade contractors, including masonry contractors, account for fifty-three percent of all construction fatalities, reports the Occupational Safety and Health Administration (OSHA). Falls accounted for the greatest number of deaths. Because of this high fatality rate, OSHA conducted seventy-one percent of its construction inspections in specialty trades in 1990.

Another danger comes from the many trades working together on a job site. A bricklayer may not be aware of a dangerous situation occurring near him because it is not related to what he is doing.

In 1986, accidents happened more often and caused more lost workdays for masonry, stonework, and plaster work than for the construction industry at large and for all private industry nationwide. See the table below.

For all the construction industry, vehicle accidents account for twenty-nine percent of deaths. Sprains and strains, especially back injuries, were the most common nonfatal injuries. Fractures were next, followed by bruises and cuts.

One-third of all masonry related OSHA citations are related to the scaffold. At least ten percent of all citations were because workers did not wear correct protective gear.

However, these scary situations should not make you shy away from construction work. Training and awareness go a long way in keeping you safe. If you take the time to learn how to run equipment the correct way on the job site, you will likely stay safe.

People most often get hurt when they are not following prescribed safety procedures, such as forgetting to wear safety goggles when using a power saw or just forgetting to be aware of their surroundings.

OSHA, a government agency formed in 1970, watchdogs work places to ensure worker safety. Much skepticism exists on the administration's usefulness and practicality, but being aware of

Accidents and Lost Workdays (1986)

	Injury or death per 100 workers	*Lost workdays per 100 workers*
Masonry, stonework, plaster	17.2	175.0
All construction	15.2	134.5
Private industry	7.9	65.8

Source: National Bureau of Labor Statistics, 911 Walnut, Suite 1604, Kansas City, MO 64106 (816-426-2481)

its regulations does go a long way in keeping job sites safer. OSHA also protects workers from contractors who might want to cut corners to save money instead of practicing safe methods.

If you are hurt on the job, every state and the District of Columbia and Puerto Rico provide unemployment compensation benefits for people who are injured on the job and disabled for any length of time. These benefits vary across the nation, from good to poor. You may also have disability insurance benefits through your employer or union.

Posters and warning signs are becoming a regular part of the job site. Safety is becoming a critical element of every construction job. Some contractors are even hiring staff to act as safety managers to ensure safe practices on the job site. These safety managers ensure the scaffold is set up properly and that masons are dressed according to the tasks they will be doing, and hold regular "lunch-box" meetings with the crew to keep everyone notified about hazards.

CHAPTER SEVEN

IS MASONRY RIGHT FOR YOU?

MASONS HAVE A LOT OF PRIDE

Successful, happy masons like to use their heads while they are using their hands. They take lasting pride in their work because it stands as a monument to their skill and dedication.

Masonry appeals to people who have the ability and desire to work independently. Fully trained masonry craftsmen often work unsupervised. They are listened to because their opinions and skills are valued. Architects and engineers do not always have the job experience to make their ideas work and they sometimes need the experienced opinion of a mason.

MASONS HAVE GOOD MATH SKILLS

A high school diploma is not required to be a mason, but it is required by union apprenticeship programs to become a journeyman mason. A high school diploma also gives you the math and problem solving skills needed to begin a masonry career.

Knowledge in math is essential because you will use it every day on the job. Masons need to know how to reduce fractions to their lowest terms and how to multiply and divide numbers with decimal points. For example, masons want to cut as few bricks as possible on the job because it is time consuming and if not done perfectly the first time, the brick is wasted and you have to cut another one. By having math skills, masons can figure out the best way to lay the brick so fewer cuts are needed.

Reading and writing skills may not at first appear important when you are in a skilled trade, but in fact they are essential. You must be able to understand instructions on packaging, write reports for insurance, and keep logs for material breakage.

MASONS PREFER WORKING OUTDOORS

The career of bricklaying is not for those who like sitting at a desk. You must like to work outside with tools and your hands. Details must be important to you. You will need manual dexterity and stamina.

People choose this industry because they prefer to work outdoors rather than in a factory assembly line. It pays better and is more rewarding. You can measure your work daily as a wall goes up.

Being a mason—whether you are a bricklayer, tile layer, or cement mason—is an exacting job. You must be precise, accurate, and neat. The kind of person best qualified for this work must have an aptitude for accurate work and have a good eye.

MOST MASONS LIVE IN OR NEAR A CITY

Masons mostly live in or around big cities, which is where most of the work is. If you want to live in a rural area, consider stonemasonry or cement masonry. About one in four masons are self-employed, doing specialized or smaller jobs such as chimneys, retaining walls, fireplaces, or tuckpointing.

Bricklayers with leadership/management skills become foremen and then supervisors. Supervisors can advance to management or become job estimators, who figure the costs of a project. Some even move on to start their own contracting business.

BECOMING A CONTRACTOR REQUIRES AN AGGRESSIVE PERSONALITY AND HARD WORK

Becoming a successful contractor takes an aggressive, hardworking personality. You will need a lot of self-discipline to do the work it takes to make a living and be successful.

It is expensive to start a business as a masonry contractor. When you do a job, you usually have to pay all expenses up front and the client (an owner or another contractor) may not pay their final installment of your bill for a long time after the job is done. You will be floating a lot of money. Many times, contractors get their start from a bank loan or someone in the industry who has faith in them.

It pays to know people in the industry for job contacts too. Being part of local associations helps you publicize yourself and your services. The masonry industry has local associations in most larger cities. Many national associations are also available to help you get information, do a better job, and to publicize your work.

MASONS MUST BE FIT AND HEALTHY

It is difficult to adapt a bricklaying career to physical disabilities, but not impossible. Masonry is strenuous work, requiring a strong back and strong hands. Contact your state office of vocational rehabilitation for more information.

CHAPTER EIGHT

FUTURE GROWTH OF INDUSTRY

Where are most of the new jobs in this country and what kind of companies really start most of the new jobs? The answer is small business and industry. In addition, masonry, stonework, and plastering are the second fastest growing small businesses according to the United States Small Business Administration.

The costs of construction industry projects total about ten percent of the gross national product—more than eighty billion dollars a year. And that number is bound to go up in the next decade because many infrastructure facilities, such as roads, bridges, sewers, and public buildings (hospitals, schools, civic centers), have reached their life spans and need repair or replacement. Building for the growing populations in this country and in the world will always be essential. And repair is especially needed because many cities' coffers cannot take the expense of building new facilities.

Of the Brick Institute of America's (BIA) $1.5-billion budget, $400,000 is for marketing and $400,000 is allocated for engineering and research. That shows that the association is banking on growth. Not only will marketing increase the use of brick, thus more jobs for bricklayers, but marketing helps increase the perceived status of the masonry construction workers and the industry as a whole. You might not think that spending money on masonry research is a smart idea; you might think that bricklaying has not changed in centuries. But that is not true. Yes, bricklayers still put one brick atop another, but the equipment and materials they use to build walls have changed. These changes have been for the better, making masonry jobs easier and the end products of a higher quality that last longer.

WHY GET TRAINING AFTER HIGH SCHOOL?

What percentage of jobs in this country require additional training beyond a high school degree but do not require a bachelor's degree?

The answer is eighty percent of the jobs. And that will be true through the year 2000, says Joyce Winterton, executive director of the National Council on Vocational Education.

From these statistics you can see that training after high school is crucial to employment success. About twenty-five percent of high school students go on to college, and of those only half graduate with a degree. But that is only one way to get training after high school. Another more common, and less expensive, route to training is through an apprenticeship or pre-apprentice program through a union or a group such as the Job Corps.

People who complete an apprenticeship program earn about fifty-two percent more than people without training after high school. That is a significant amount that leads to a more comfortable lifestyle and an ability to earn enough money to have the extras, like money for vacations, your children's education, or even a fancy car, if that is your pleasure.

THERE IS A SHORTAGE OF MASONS

Every healthy industry needs new people coming into it. However, the masonry industry needs new people even more than most other industries. The average age of a bricklayer is fifty-three, according to the Department of Labor. That is close to retirement age. Within ten years about half the current bricklayers will have retired. That means there is plenty of room for you in the masonry industry. You will not have to worry about a glut in the market of skilled masons. Even the computer industry, once thought to always provide employment, is experiencing a glut of workers. The building industry may have highs and lows, but it will always exist as long as people do. Everyone needs a place to live, to work, to worship, to play, and to learn.

Brick Institute of America (BIA) president Nelson Cooney says another 55,000 or more bricklayers are needed to meet current demand and another 26,000 by the year 2000. John A. Heslip, president of the National Concrete Masonry Association (NCMA), says attrition is producing about a 6,000 mason shortfall each year. We are going to need 7,000 masons annually to keep up with growing demand and attrition. In 1992, Bill Weaver, BIA education director, also said that more than 7,000 new masons are needed for the next ten years.

These numbers show that at least three percent of the total number of union and nonunion masons need to be replaced each year just to keep up with those leaving the industry through retirement, a promotion in the field, or new careers.

In 1988, the union had 903 programs with 4,446 apprentices in training. With the attrition rate at more than 6,000 a year and only about 4,500 coming in, the mason shortfall is somewhere between 1,500 and 2,000 annually just to maintain the present work force size. And since the need for trained masons is currently expanding, many more are needed.

The masonry industry will be needing 130,000 to 140,000 more trained masons in the next ten years. In March 1988, about 160,000 union and nonunion masons were in the work force. That number was up to over 200,000 in 1992.

At the 1989 convention of the Masonry Contractors Association of America (MCAA), C. DeWitt "Dee" Brown, Jr., of Dee Brown Masonry, Inc., Dallas, Texas, said: "According to a recent survey, we will be importing skilled craftsmen if we do not do something dramatic between now and 1994."

	Estimated Employment, 1986	Projected Employment, 2000	Percent Change, 1986-2000	Annual Average Openings
Bricklayers, Stone masons MODERATE GROWTH	161,000	187,000	+16.1%	1.214
Cement masons Terrazzo workers RAPID GROWTH	118,000	142,000	+20.3%	1,714
Plasterers SLOW GROWTH	28,000	31,000	+10.7%	214
Tilesetters MODERATE GROWTH	32,000	39,000	+21.9%	500

Source: Dept. of Labor

The International Union of Bricklayers & Allied Craftsmen (BAC) reported on the Department of Labor statistics for job growth of bricklayers:

In 1990, there were about 152,000 bricklayers and stonelayers. By 2005, the Department of Labor says this country will need 183,000 bricklayers. That's an average annual increase of 1.2%.

WHERE ARE BRICKLAYING JOBS IN THIS COUNTRY?

Bricklayer shortages tend to be regional, not national. In 1988, Boston was begging for more bricklayers. Houston was not building so some bricklayers were unemployed. In 1989, Neil Humphries of Boren Clay Products said there was a shortage of about one thousand five hundred bricklayers in South Carolina.

The need for bricklayers in different parts of the country varies. For example, the BAC estimates that from 1990 to 1996, there will be a high demand in the following states:

BRICKLAYER GROWTH

State	from 1990 to 1996
Colorado	29.4%
Nevada	34.0%
Utah	12.1%
New York	8.8%
New Jersey	4.3%

STONEMASON GROWTH

State	from 1990 to 1996
Colorado	28.4%
Nevada	Not available
Utah	15.2%
New York	Not available
New Jersey	0.0%

CEMENT MASONS (FINISHERS) GROWTH

State	from 1990 to 1996
Colorado	31.7%
Nevada	46.0%
Utah	19.0%
New York	Not available
New Jersey	2.1%

These numbers do not take into account the demand that is created by people exiting the trades or retiring.

In Illinois, 1986 employment was at 14,424, with a projected need for 3,751 more masons by 2000. That is a twenty-six percent growth. Similar state data show that the growth from 1990 to 1996 will be almost ten percent. The construction trades are enjoying a larger percentage of growth than any other field rated by the Department of Labor.

THE UNION HELPS YOU FIND WHERE THE JOBS ARE

To keep masons employed, the International Union of Bricklayers & Allied Craftsmen (BAC) has a program called Job Information Center (JIC). BAC members who are willing to travel complete a three-page enrollment form. This form asks them to describe their skills and supervisory experience. It also asks to which states they are willing to travel. This data is entered into a computer by the job information center in St. Louis.

When a local union exhausts its local manpower resources, it requests a computer printout of qualified craftsmen from the Job Information Center. From this list, the local union then calls a mason from the list that it might want to hire. If you are a union mason and able to travel to work on a job site for a period of time, you are almost always guaranteed work through this program. Again, the shortage of masons depends on where the building is taking place.

BRICK USE IS FAVORED

Competition from other building materials and systems is increasing. But more brick homes (veneer mostly) are built now than thirty years ago because of the trend toward easy maintenance, an effort to lower insurance rates, and beauty. These homes are mostly partial or full veneer brick on a wood frame. Few homes today are built of structural brick (brick that holds up the structure). As long as brick is a favored material, you will always have a job. And brick has always been favored.

A National Association of Home Builders' survey shows an increase of brick used as an exterior material from about twenty percent in 1940 to thirty percent in the mid-1970s. Bricks used in combination with another material covers about a third of all houses. Wood is second at twenty-eight percent and aluminum is third at about ten percent.

Of the 131 architects responding to a 1990 survey by the Masonry Institute of St. Louis, eighty-nine percent choose masonry because of durability, seventy-nine percent because of availability, seventy-five percent for texture, and seventy-three percent for good value. A majority of eighty-three percent of them believe that future architectural styles will favor masonry. Facts like these show that masonry will continue to be the desired building finish. That means jobs will continue to be plentiful in building with it.

MASONRY UNIT MANUFACTURING IS CONSOLIDATING

The entire masonry industry now has $16.5 billion in annual sales, says John Heslip, 1989 president of National Concrete Masonry Association (NCMA). That is only a small share of the total construction dollars spent in this country, but it still makes for a lot of building and a lot of jobs. The following are descriptions of various facets of the industry.

Brick Manufacturing Industry: In 1990, there were 112 brick manufacturers with 229 plants in forty states that produce ten billion eight-inch units each year. About 35,000 people are employed to produce brick and another 15,000 employed with dealers and distributors of brick. Those are jobs that do not necessarily require you to have been a mason first. Total annual wholesale value of brick is one billion dollars.

To produce brick in modernized plants is not an artisan craft, although if you were making bricks by hand, you would be an artisan. Making bricks in plants takes a knowledge of how to run the equipment and other machines involved in making, firing, transporting, and storing the brick. It is more like a factory worker's position. You work both inside and outside, however, and you produce something of value to everyone.

Future Growth of Industry 53

In 1942, there were about 4,000 to 5,000 brick plants in the United States. Only thirty years ago there were 3,000 brick manufacturers. Of the current 112 manufacturers, fifty-five percent of all brick is produced by only ten companies. Just over a third of all brick production used in the United States is done outside the country. You can see that the industry is being run by fewer and fewer companies, but each is producing a colossal quantity of brick.

Concrete Masonry Industry: Thirty years ago there were 5,000 companies making concrete masonry units. In 1985, there were 1,011 block producers; in 1988 there were 980 producers. By 1990, there were 925 manufacturers of concrete masonry with 1,450 plants all over the United States. Like the brick manufacturers, the concrete block manufacturing industry is also consolidating.

Sales of concrete block are up, so each manufacturer is producing more and more. When sales of brick and block are up, that means jobs for masons are increasing as well, because someone has to put all these materials in place.

It's estimated that 5.5 billion eight-inch equivalent units are produced each year. In the last ten years, sales have increased by twenty-five percent. Total direct employment in the block industry is 55,000. Those are the people who manufacture the block at the plants, run the manufacturing businesses, and make the product sales. Total annual sales for the concrete block industry are $3.8 billion.

Market share of concrete block was 1.65% of total construction dollars spent in 1986 (80.3% of that 1.65% was common block, 10.5% was architectural block). Sales of architectural block with textures or colors is growing steadily. More and more architects are learning that they can build relatively inexpensively, yet maintain good design by using architectural concrete block. Check out many of the newly built discount retail stores; they often are built of architectural concrete block.

Stone Industry: Total annual installed cost of stone in the United States is about $2.5 billion. The stone industry is the most fragmented segment of the masonry industry. In the United States there are only six major fabricators, but there are hundreds of small, family-owned businesses.

About ninety percent of natural stone firms are members of Marble Institute of America, Building Stone Institute, National Building Granite Quarries Association, or the Indiana Limestone Institute.

Pavers Industry: Look around the fancy neighborhoods and you will see a lot of driveways with pavers. It has become the decorator look of the 1980s and 1990s. They offer color, pattern, durability, and long life for driveways, patios, and sidewalks in many cities across the nation. They are made of either brick or concrete. The same manufacturers who make brick and block usually also make pavers.

CHAPTER NINE
UNION VERSUS NONUNION

A young person can enter the industry in many ways. But by far the most common doorway into masonry is through craft training known as an apprenticeship.

The construction industry is generally unionized, except for some segments of the residential market and some geographic areas. Labor unions represent workers so that the workers can achieve good working conditions, good benefits, and good wages.

Nonunion bricklayers can make more than union bricklayers, but this is not usually the case. Nonunion bricklayers also usually do not have the same guaranteed benefits that union bricklayers enjoy. However, nonunion bricklayers do not have to pay part of their hourly wage nor dues to any union.

There are more than 120,000 union bricklayers in the United States. The number of nonunion bricklayers is probably comparable, but figures are not definite for nonunion workers because it is harder to count them. They do not all belong to a group where they can be counted and many sift in and out of the industry.

The institutional union type structure, or the guild structure, started in Europe and moved to this country in the early part of the century. It continued into the 1930s when it became more formalized under the Fitzgerald Act and the Apprenticeship Program was established.

Unions for people in the masonry industry include: the International Union of Bricklayers and Allied Craftsmen (BAC), Laborers International Union of North America, the Operative Plasterers' and Cement Masons' International Association of the United States and Canada, and the United Cement, Lime, Gypsum, and Allied Workers International Union.

Many associations and other organizations represent the industry. See Appendix E for an extensive list.

DO I GO UNION OR NONUNION?

The training programs to become a mason differ between union and nonunion. The training route you take will determine whether you are union or nonunion. Most people in the industry would recommend going the union route. But as more and better nonunion training programs have been started, the line between the quality of work provided by union and nonunion is starting to blur.

Strong feelings exist about whether to use union or nonunion workers on those few occasions that it is an option. It is often less expensive for a contractor or owner to use nonunion workers. Many areas have requirements as to whether or not you can hire nonunion workers.

At the twentieth annual convention of the International Masonry Institute (IMI) in November 1990, John Joyce, president of the International Union of Bricklayers and Allied Craftsmen (BAC) said, nonunion competition will continue as long as "clients of the building industry continue to delude themselves into believing that nonunion construction is a bargain."

General contractors who hire masonry subcontractors prefer the subcontractors bidding for the masonry segment of a job to be all union or all nonunion. That way a fair comparison can be made to determine which contractor has the best prices and what quality of work will be performed.

Union wages in any one area are all the same. For example, all union bricklayers in Chicago make the same hourly wage. It does not matter how long you have been a journeyman mason or how old you are, or what your sex or race is. If you are a journeyman mason, you are assured that every other mason on the wall with you is earning exactly the same amount.

EDUCATING MASONS

Relationships between the union and many masonry associations have been strained over the years because of disagreements on how a mason should be trained. Unions say that any training that is not a union apprenticeship is inferior training to theirs. Nonunion training program leaders say that their programs are at least as good as the union training programs. Nonunion training has expanded greatly in the last five years.

To learn to become a skilled mason there are four routes of training: apprenticeships, open-shop training, vocational educational liaison, and curriculum development. Most people enter the masonry industry through union apprenticeships. Many people become skilled workers without formal training, though, by starting as a laborer and learning on the job in a nonstructured way. Others enter a nonunion training program set up by the Job Corps or another industry group.

The union apprenticeship program is firmly entrenched in this country and produces skilled journeyman masons. The BAC union says that most vocational programs do not produce bricklayers that can do quality work, and that this harms industry credibility.

The BAC union says vocational training is not the answer to the bricklayer shortage. It is confident that its apprenticeship program meets the training needs of young people and the industry. The union says the bricklayer shortage is only regional; if bricklayers were moved to where the jobs are, construction needs would be met.

In March 1988, John Heffner, associate director of Manpower and Training Services of the Associated General Contractors of America, said that unfortunately contractors frequently do not see the need to train bricklayers. These contractors say they are still able to hire unemployed union bricklayers. But this must not be true because, at a negotiating conference attended by sixty contractors, the biggest concern they expressed was how to find bricklayers to do the work. Several contractors refuse to bid on masonry jobs because they cannot get the bricklayers to complete them. It appears that a mason shortage exists, but few contractors want to be bothered with training more of them.

This is a tragedy of the system. To become a union apprentice you must work for a contractor. But if few contractors are willing to train beginners then there will be fewer masons available overall. This is where the Job Corps and other nonunion training programs run by the government in conjunction with local masonry related associations step in.

The nonunion sector did not have training programs available until about 1982. It is difficult to get a nonunion program established, but many do exist. Chapters Eleven and Twelve describe some of the available programs.

RESIDENTIAL MASONRY IS USUALLY BEST FOR NONUNION MASONS

There are far more workers employed laying brick and block in residential areas than there are in the commercial and industrial markets, where union journeymen are usually used.

To build brick veneer houses, you do not need the same skill level as a journeyman bricklayer. All the brick are simply laid to a line, including the corners. Likewise, it does not require a journeyman skill level for a mason to build concrete block basements.

That is not to say that residential work is dull or simplistic. Many residences still require a masonry fireplace and you need to be highly skilled to build a full masonry fireplace. Oftentimes in residences, the brick is positioned in a pattern for a pleasing look. Knowing the mathematics of getting brick set in a pattern with no loose ends is an art.

CHAPTER TEN

UNION APPRENTICESHIPS

WHAT IS AN APPRENTICESHIP?

An apprenticeship is a formal training method to learn a craft or trade using on-the-job training with classroom work. After spending the prescribed amount of hours or years in an apprenticeship program, you receive the credentials of the sought-after title journeyman.

An advantage to being an apprentice straight out of high school is that you are paid while you learn. You are a student and you are working on the job. Although you often start out at half the wage of a journeyman, you still are making a lot more than the twenty-five percent of high school graduates who are unemployed when they first come from school.

WHY HAVE AN APPRENTICESHIP?

Over the centuries most trades and specialty industries have had training programs so that new workers could learn from the experienced workers. For example, doctors have to do residencies before they are complete in their training. In the same way, apprentices practice as amateurs before they join the rank of journeyman.

Having this training period before becoming a journeyman helps you perfect your skill and gives you more credibility as a mason. It also makes your work more valuable to the contractor.

WHO DOES AN APPRENTICE WORK FOR?

Apprenticeship programs are cooperative efforts of the union and employers (contractors). The contractor you work for pays you for your time. The wage is a fixed amount, based on the collective bargaining agreement between the union and the area contractors. As an apprentice you will be given a variety of tasks.

These traditional apprenticeship programs focus almost exclusively on newcomers to the trade. Journeymen in one branch of the trade have little opportunity to learn other specialties, and supervisory personnel have to seek further training outside the union training programs, usually through vocational supervisory classes at local junior colleges, seminars through trade shows, or training programs run by masonry industry associations.

WHO PAYS FOR THIS APPRENTICESHIP TRAINING?

Training skilled masonry craftsmen is the International Masonry Institute's largest activity. In 1990, IMI spent more than four million dollars on apprentice and journeyman training. That is less than one-sixth percent of the masonry industry's annual sales and much less that what other industries spend on training. Eventually the institute plans to spend twenty million dollars a year on training.

In the IMI local classrooms you learn blueprint reading; trade math; masonry materials, tools and equipment; and safety practices. On the job each day, you learn the daily skills of becoming a journeyman mason.

Nine branches of the masonry trades are covered in the national IMI training program: bricklaying, tile laying, stone masonry, plastering, cement masonry, marble masonry, building restoration and maintenance, terrazzo, and mosaic work.

Historically, labor/management training programs in the masonry industry have been run at the local and regional level. These programs are administered by Joint Apprenticeship and Training Committees (JATCs) composed of contractors and union representatives. The programs vary from locale to locale, and some are better funded than others. In total, local programs spend about seven million dollars annually.

As an apprentice, you probably have to pay initiation fees and a very small amount of your hourly wage to the union. These amounts vary between local unions. For example, in southern California, to become an apprentice requires you to pay an initiation fee of $200.00 ($100.00 down and $100.00 after sixty days). Then you pay another $200.00 upon completion of the apprenticeship. This money goes to the union.

UNION BRICKLAYING TRAINING IS CONSISTENT ACROSS THE NATION

In 1986, IMI launched a centralized, national Trowel Trades Training System designed to serve contractors and workers in all branches of masonry, at all levels. It offers these benefits to the masonry industry:

- national standards for both the technical knowledge and job skills of masonry craftsmen
- improves the portability of worker skills, helping the industry deal with geographic variations on labor supply

- assures a high quality, consistent level of training both from region to region and from year to year
- helps masonry solidify and expand its market position

This national training system does not replace the JATC-based apprenticeship system. JATCs that have proved their effectiveness in local work force planning are a vital part of the new program.

Apprentices go through a pre-job training program, then have several years of on-the-job training before becoming journeymen masons. In this new system, each apprentice receives up to twelve weeks of training before being on the job for the first time. This enables apprentices to earn their way so contractors do not have to subsidize their on-the-job training. Then apprentices enter a three-year program that combines schooling and on-the-job experience.

Backing the nationwide system is a headquarters data management system, a computerized record-keeping system that tracks all individual training, qualifications, and other details. This system provides manpower planning support and job referrals throughout the industry.

An international training center and several regional centers are planned throughout the United States and Canada. In the meantime, leased facilities in Piney Point, Maryland, are used for instructor training, and rented facilities in Boston, New York City, Chicago, Detroit, and Los Angeles are used for regional and local training programs.

National standards have been developed by the Bureau of Apprenticeship and Training in consultation with international unions. These programs are coordinated with local joint labor-management committees that also develop admission standards. Generally such standards require that an applicant be between nineteen and twenty-five years old and have from nine to twelve years of education. These standards are not rigid, nor do they apply nationally.

WHERE DO I GET APPRENTICESHIP INFORMATION?

Detailed information on apprenticeship can be obtained from the individual trade unions and from various state and federal apprenticeship agencies (appendix B). Each state has many union offices. Illinois, for example, has more than thirty bricklayer union offices. Call your local Department of Labor for addresses and phone numbers.

If you were to start your apprenticeship in southern California, in May 1993, you start at $10.55 an hour, plus benefits. By the last six months of your three-year program you will be earning $23.73 an hour. Apprentices in southern California must be at least 17 years old. In Chicago, they must be at least 18 years old. Check with your local union for specific requirements.

Most masons complete a three-year apprenticeship program that consists of about 6,000 hours of on-the-job training and classroom training. Half that time is spent laying bricks; 600 hours in laying stones; 200 hours in pointing, cleaning, and caulking; 300 hours in fireproofing; 1,700 hours in terra-cotta, tile block, concrete block, and artificial (concrete) stone. The apprentice becomes proficient enough to read a blueprint and build from it.

HOW DO I BECOME AN APPRENTICE?

To become a bricklayer union apprentice, you must apply at one of your local bricklayer union offices. They are listed in the phone book under "Bricklayer Union." Each has slightly different requirements. For example, in the Local 21 union in Chicago, to enter the three-year apprenticeship program you must have the following:

- A high school diploma or GED certificate
- Transcripts of at least eight high school credits

- An aptitude test and an interview
- A physical exam
- A birth certificate
- Three references
- Be at least 18 years old

All bricklaying and other trowel trades apprenticeship programs offer equal opportunity to all people, regardless of race, age, color, sex, religion, or nationality.

Responsibilities and obligations of the apprentice are as follows, as recommended by the United States Department of Labor:

1. To perform diligently and faithfully the work of the trade and other pertinent duties as assigned by the contractor in accordance with the provisions of the standards.
2. To take care of materials and equipment and abide by the working rules of the Contractor and the local Joint Apprenticeship Committee.
3. To attend regularly and complete satisfactorily the required hours of instruction in subjects related to the trade, as provided under the local standards.
4. To maintain such records of work experience and training received on the job and in related instruction as may be required by the local joint committee.
5. To develop safe working habits and conduct themselves in such manner as to assure their own safety and that of their co-workers.
6. To work for the contractor to whom assigned to the completion of the apprenticeship, unless the apprentice is reassigned to another contractor or the agreement is terminated by the local joint committee.
7. To conduct themselves at all times in a creditable, ethical, and moral manner, realizing that much time, money, and effort are spent to afford them an opportunity to become skilled craft workers.

8. To abide by the work rules of the Collective Bargaining Agreement negotiated and administered by the local Joint Apprenticeship Committee.

WHAT WILL I LEARN IN AN APPRENTICESHIP PROGRAM?

The following course outline is recommended by the United States Department of Labor and is generally followed for the bricklayer apprenticeship. Every six months the apprentice advances to the next level.

FIRST AND SECOND YEARS

Masonry Materials

1. masonry units
   history, description, manufacture, classification, types, special units, structural characteristics, physical properties, color, texture, and uses for:
   a. clay and shale brick
   b. fire brick
   c. sand-lime brick
   d. concrete masonry units
   e. tile (structural and facing)
   f. stone (granite, limestone, sandstone, marble)
   g. acid brick
   h. glass block
   i. terra-cotta
2. mortar
   properties, description, uses, workability, water retentiveness, bond, durability, and admixtures for:
   a. hydrated lime
   b. cement lime
   c. cement mortar

d. prepared masonry cement mortar
   e. special mortars for:
      (1) firebrick
      (2) glass block
      (3) acid brick
      (4) stone (granite, limestone, sandstone, marble)
3. sand
   classification, description, selection, tests, types, and uses

Tools and Equipment
Use, care, operation, and safe practices for:

1. brick trowel
2. brick hammer, blocking chisels, six-foot rules, levels, and jointing tools.
3. story pole and spacing rule:
4. stone setting:
   a. woodwedges
   b. setting tools
   c. caulking gun
   d. chain hoists
   e. cranes
   f. hangers
5. accessories:
   a. wall ties
   b. expansion strips
   c. clip and angles
   d. nailing blocks
   e. reinforced steel for grouted walls and lintels
   f. steel and precast lintels
   g. flashing materials
   h. anchor bolts
   i. steel-bearing plates
6. welding equipment

Trade Arithmetic

1. review the fundamental operations of arithmetic, including:
    a. fractions
    b. decimals
    c. conversions
    d. weights
    e. measures
2. reading the rule
    a. six-foot rule
    b. spacing rule

Plan Reading, Blueprint Reading, and Trade Sketching

1. fundamentals of plan and blueprint reading:
    a. types of plans
    b. kinds of plans
    c. conventions
    d. symbols
    e. scale representation
    f. dimensions
2. trade sketching:
    a. tools (types)
    b. straight-line sketching
    c. circles and arcs
    d. making a working sketch

Construction Details

1. trade terms, motion study, bonds (structural and pattern), laying of units, joints, and so forth, for:
    a. walls
    b. footings
    c. pilasters, columns, and piers
    d. chases
    e. recesses (corbelling)
    f. chimneys and fireplaces

2. cleaning, caulking, and pointing
3. reinforced masonry lintels

Shop Practices
1. spreading mortar
2. laying brick to line (building inside and outside corners for a four-, eight-, and twelve-inch wall)
3. layout and erect:
    a. walls and corners with:
        (1) Flemish and Dutch bond
        (2) tile backing
        (3) pilasters and chase
        (4) a cavity
        (5) reinforced grouted brick
    b. brick piers
    c. chimneys (single and double flues)
4. setting sills, coping, and quoins
5. laying paving brick

Safety

THIRD YEAR

Tools and Equipment
use, care, operation, and safe practice for:

1. building's level and transit
2. frames, beams, lintels, and rods
3. welding equipment

Blueprint Reading
1. specifications
2. job layout
3. shop drawings
4. modular measure

Construction Details
1. arch construction
2. modular masonry
3. firebox construction
4. layout of story poles and batter boards

Estimating
1. mortar
2. masonry units (modular and non-modular types)
3. concrete footings

Shop Practice
Lay out and erect:
1. reinforced masonry lintels
2. story pole and batter boards
3. fireplaces (with and without steel fireplace forms)
4. project with glazed tile leads and panels
5. project with marble or granite setting, adhesive, terra-cotta, glass block
6. modular wall
7. circular corner

Prefabricated Masonry Panels
1. layout
2. assembly
3. welding and erection
4. installation
5. caulking, pointing, and cleaning

Applications of Insulating Materials for Masonry Walls
1. theory
2. care and preparation of area
3. types of application

Safety
1. valid safety certificate
2. valid first-aid certificate

Here is how the hours of the apprenticeship program break down:

Work experience	*Hours*
laying of masonry units	3,000
laying of stone	450

1. cutting and setting of rubblework or stonework
2. setting of cut-stone trimmings
3. butting ashlar

pointing, cleaning, and caulking 150

1. pointing brick and stone, butting and raking joints
2. cleaning stone, brick, and tile (water, acid, sandblast)
3. caulking stone, brick and glass block

installation of building units 525

1. tile cutting and setting
2. cutting, setting, and pointing of special masonry units
3. blockarching
4. mixing mortar, cement, and patent mortar; spreading mortar; bonding and typing
5. building footings and foundations
6. plain exterior brickwork (striaght wall work, backing up brickwork)
7. building arches, quoins, columns, piers, and corners
8. planning and building chimneys, fireplaces flues, and floors and stairs
9. building masonry panels
10. laying paving brick

fireproofing 225
1. building party walls (partition tile, gypsum blocks, glazed tile)
2. standardized firebrick
3. specialties
care and use of tools and equipment 150
1. trowels
2. brick hammer
3. plumb rule
4. scaffolds
5. cutting saws
6. welding equipment
TOTAL 4,500

Part of this 4,500 hours are 144 hours each year of related instruction, usually in a classroom. This instruction covers theory, terminology, and general trade information and practices, including blueprint reading and safety.

After satisfactorily completing 4,500 hours work (generally three years), the apprentice receives a Certificate of Completion of Apprenticeship, which means the apprentice can join a union to become a fully qualified journeyman—a mason.

OTHER TROWEL TRADE APPRENTICESHIPS

To become a journeyman in cement masonry, tile setting, stone masonry, or plastering you must go through an apprenticeship program that is similar to the way the bricklaying apprenticeship program is run. Pay varies for apprentices in different unions. You generally have to be employed with a contractor as a sponsor to become an apprentice through the cement union.

TILE LAYING AND TERRAZZO APPRENTICESHIP

The tile laying/terrazzo apprenticeship lasts three years. Before you are accepted you must pass an aptitude test, an oral test, and a physical examination. Then you must pass a pre-apprentice program. In Chicago, that program lasts eight weeks for eight hours a day. Then the union finds you a position with a contractor and you are indentured to the union.

While you work for a contractor as an apprentice, you are paid a percentage of the journeyman's wages. For example, in Chicago, you start at 50 percent of a journeyman's wages. After every 1,000 hours of work (about six months) your wages go up ten percent. During the apprenticeship you must attend 144 hours of class time a year, which might be every other Saturday.

CEMENT MASONRY APPRENTICESHIP

In most cases you must be employed with a contractor to be considered for a cement mason apprenticeship and be at least eighteen years of age and have received your high school diploma or passed the GED. The age was recently raised from seventeen to eighteen because of federal insurance requirements. A physical examination including a drug test is required.

The contractor you work for sponsors your apprenticeship training. The training program you follow on the job site is determined by the contractor.

The union supplies the classroom training. It usually consists of one night a week for a three-hour class. In Chicago, for example, each class lasts thirty weeks and each covers subjects such as:

- using a level and transit to keep work level and pitch correct
- reading blueprints

- using a carpenter's rule (in 16ths) and an engineer's rule (in 100ths)
- welding screeds for bridgework
- safety
- estimating concrete quantities

It is critical that the cement mason can read and write well. Manufacturer's instructions often need to be followed exactly for concrete additives to produce the desired and specified results. If you do not follow the written instructions on a package, you are liable for the mistake and can cost your business a lot of money when concrete needs to be ripped out and replaced.

A strong math background is needed as well. Keeping a floor level, translating measurements between different types of rules, and ordering the right amount of concrete all take math skills.

MARBLE SETTING APPRENTICESHIP

Most often, marble setting apprenticeships are affiliated with the local brickmasonry unions. There are only two union offices that are solely for marble setters. One is in New York and the other is in Chicago. The apprenticeship for a marble setter takes three years.

The marble setting union in Chicago has discontinued its apprenticeship training program because the number of jobs available in the Chicago area for stonemasons is not enough to merit further training by that union at this time. Its last apprenticeship program was in 1987. Only fifteen youths were apprenticed and of that number, only six found jobs as marble setters.

Of the 125 members of Chicago's marble setters union, 119 are working every day, which is a good rate. In 1992, the International Masonry Institute (IMI) started looking into setting up classrooms for marble setting in different areas of the country where there were jobs.

Marble setting career opportunities have slowed down in general across the nation. In the early 1960s, marble setting was at its height when many people and businesses could afford the luxury of marble.

CHAPTER ELEVEN
NONUNION TRAINING

One way young people can acquire bricklaying skills is by entering a pre-apprenticeship, occupational training program. High schools, vocational schools, community colleges, and local masonry groups offer such programs.

In a March 1988 speech, Cecil Simpson, a manager at Manpower Development, division of the National Concrete Masonry Association (NCMA) said that secondary schools enroll about 20,000 students in masonry programs across the United States. Also, there are about 5,000 apprentices in the mason training programs nationwide.

PRE-APPRENTICESHIP PROGRAM
PROVIDES BASIC SKILLS

After satisfactorily completing a pre-apprenticeship training program, a prospective mason should have enough marketable skills to be hired by a contractor but he or she will not be a fully trained mason. The person will begin work as a laborer, a mason tender, or an apprentice.

Pre-apprenticeship programs give you the basic skills you need to enter an apprenticeship program. Some of these programs, such as the Job Corps programs described in the next chapter, also provide high school dropouts with the knowledge they need to complete the GED, which is comparable to a high school diploma.

Some people take an unskilled labor position with a contractor to await an apprenticeship position opening. This work provides a good learning experience, but it does not prepare you to be a journeyman mason. It may help you decide to enter masonry or one of the other skilled trades.

Some programs are more comprehensive than others. Some just provide the basic skills for a lower level construction position and others provide the level of skills that a union journeyman has. Even though your skill level may match a union journeyman's, you cannot be hired by a union contractor until you go through the three-year apprenticeship program. As a nonunion mason, though, you can sell your skills anywhere that does not require union masons.

Any bricklaying curriculum should provide math, science, English, and writing instruction. Useful courses in high school include algebra and geometry. Trade courses such as architectural drawing and carpentry are also helpful. To build with masonry you need both knowledge and skill. The tools, equipment, materials, and safety concerns must all be learned.

There are two types of masonry construction: conventional and engineered. Engineered masonry construction takes much training to attain the knowledge necessary to make a building structurally sound. Engineered construction usually involves steel reinforcement as an integral part of the building.

To learn more about local vocational masonry training courses, contact the Vocational Industrial Clubs of America, Inc. (VICA). Founded in 1965, VICA is the nation's largest and only organization for trade, industrial, technical, and health occupation students. There are 14,000 VICA chapters in the United States and more than 276,000 students involved. See appendix C for the VICA director office in your state.

Members of VICA are usually high school students, but there are VICA branches at vocational schools and junior colleges too. The club activities are part of the students' school curriculum.

VICA has a computerized employment network that links prospective employers with prospective pre-apprentices. For more information, contact: VICA, P.O. Box 3000, Leesburg, VA 22075 (703-777-8810).

Another program, called the Wheels of Learning Program, is a three-year masonry program that is broken into seventy-two instructional unit modules, with 150 hours of prescribed instruction per year. It is run by Mary "Toni" Dimante, the director of education for the Associated Builders and Contractors Inc. (ABC) in Ohio. Wheels of Learning is being used throughout the country in more than eighty chapters of public and privately funded training programs. In 1988, 325 masonry apprentices graduated. There are more than 2,000 masonry apprentices in ABC's programs throughout the country.

SOME LOCAL NONUNION PROGRAMS

In Washington, D.C., The Masonry Institute, Inc., Bethesda, Maryland, offers bricklaying classes funded by a Private Industry Council (PIC). The Masonry Institute is a local association funded by masonry contractors and material suppliers to promote the use of masonry in the Washington, D.C., area.

The District of Columbia PIC is just one of more than 600 PICs in the United States. Made up of local private and public leaders, PICs determine the job skills needed in their communities. They then use federal funds from the Job Training Partnership Act (JTPA) to set up programs to give low income, unemployed people the skills needed to enter the local work force.

In 1987, the Masonry Institute received a $57,000 grant for two six-week classes that in 1988 provided the area with forty-eight new pre-apprentice masons. There were 160 applicants for the first class. After four weeks of training, it only lost one trainee. By 1989, it had held seven classes, and the Masonry Institute has graduated 118 people from its program.

Students spend more than seventy-five percent of their time laying brick and block. They learn job site safety, using masonry tools, mixing and spreading mortar, and using scaffolding. Math for masons, wall types, what contractors expect from bricklayers, and laying brick and block are included. Guest speakers, usually foremen, also speak to the classes.

Field trips to job sites and a "meet the contractors day" show the Institute's contractor members—the contractors with whom the students are guaranteed jobs—what the students learned in the program. Graduates can lay to the line and are employable. The mason contractors in the D.C. area guaranteed jobs to all who complete the program, beginning at $7 an hour.

Copying the success of Washington, D.C.'s program, the Masonry Institute of Indiana started a program in the spring of 1989 and trained six students for thirteen weeks. The first week was spent on general employability training and five weeks of masonry training in the classroom. The rest of the time was spent on the job. Contractors paid one-half the students' wages and the government grant paid for the rest.

The Fast Track Bricklayer Training Programs are sponsored by the Brick Association of North Carolina (BANC). Most classes are only two weeks long, but with minor changes to the curriculum they can run for three or four weeks.

In that time, students learn the basics of bricklaying, such as making mortar, laying to the line, using a corner lead and pole, plumbing a jamb, safety, and caring for and using tools. High school masonry instructors teach five eight-hour days of classes per week.

Students build actual projects such as retaining walls. Most of the students are unemployed adults in the area. BANC has held the class in eight cities. Graduates, usually fifteen per class, receive certificates and job placement assistance to apprenticeship programs.

The Maryland School for the Deaf has a masonry/construction trades program that it started in 1978. Freshman students choose two trades to study. In their sophomore year they must choose one of the trades for concentrated training over the next three years. Trainees go to one-and-a-half hours of classes five days a week during the ten-month school year. They learn to lay brick and block, mix mortar, build scaffolds, and how to be a mason tender.

Local companies donate masonry supplies for the classes. To date, students have completed fifty-nine projects—funded by private donations, school maintenance allotments, and state money—on the school campus in the last ten years. Fireplaces, brick flower beds, brick veneer walls, and brick walkways are some of the projects. Seventeen of the school's students have continued to build with masonry and become masons or contractors.

GET AN EDUCATION WHILE REBUILDING AMERICA

As part of their on-the-job training, pre-apprentice masons in various training programs restore historic structures to their original appearances. One such program, "Partners in Preservation: You're Keeping History Alive," is a partnership of the Job Corps, a training program for economically disadvantaged youth, and the National Trust for Historic Preservation.

The pilot project, a 110-year-old one-room schoolhouse called Mountain Gap School, included a variety of masonry work. Pre-apprentice masons restored the stone foundation, rebuilt the chimney, and built an exterior stone retaining wall. The Leesburg, Virginia, project was completed in September 1990 after two years of restoration work and received a county award.

Another program conducted by the IMI in 1990 gave young masons the chance to renovate a historic train station built in 1911 in Rome, New York. The project includes repointing and cleaning exterior brickwork and stone trim, weatherproofing, and restoring interior marble stairs.

TRAINING FOR CONCRETE MASONRY SALES

The Construction Specifications Institute (CSI) and the National Concrete Masonry Association (NCMA) provide technical training for concrete masonry salespeople. You can receive a certificate making you a Certified Consultant of Concrete Masonry (CCCM). For more information contact CSI, which is listed in appendix E.

HOW TO BECOME A CERTIFIED MASONRY INSPECTOR

Section 306 of the Uniform Building Code allows the use of special inspectors for certain construction elements and phases. To establish a standard of professionalism among such inspectors, the International Conference of Building Officials offers voluntary certification exams in thirteen UBC defined categories, including structural masonry.

The structural masonry exam has two parts. The first part is a two-hour, open-book, code and plan reading test and the second part, also two hours long, is closed-book and covers general knowledge. For more information: Douglas Thornburg, vice president, education, ICBO, Education, 5360 S. Workman Mill Rd., Whittier, CA 90601 (213-699-0541).

YOU CAN GET COLLEGE EDUCATION IN CONSTRUCTION

More and more colleges are offering bachelor's degrees in construction. These usually focus on materials, estimating, and management rather than hands-on applications. They are ideal for people hoping to move out of the field into supervision, management, or contracting ownership.

Even the number of construction-related graduate programs has increased this past decade. The degree offered is usually the master of science in building construction or construction management.

Clayford T. Grimm says, "The vast majority of structural engineering programs teach their graduate students virtually nothing about masonry. There probably are not more than a dozen graduate masonry engineering courses in this country. I know of only one school that has three masonry engineering courses."

A 1982 survey by The Masonry Society (TMS) showed that fifty major universities offered full or partial courses on masonry. In a 1988 TMS survey about eighty universities had such courses.

CHAPTER TWELVE

JOB CORPS TRAINING

WHAT IS THE JOB CORPS?

Job Corps is the world's oldest and largest residential vocational training and basic education program for at-risk youth between the ages of sixteen to twenty-one. It is a network of training centers across the United States which helps young people earn a GED and achieve entry-level proficiency in a trade. The program lasts at least 180 days (half a year) but usually runs longer.

The Job Corps prepares underprivileged young people and high school dropouts for trades such as cooking, forestry, plastering and cement masonry, clerical skills, carpentry, heavy equipment operations, painting, automotive trades, and bricklaying.

In 1987, eighty-four percent of the Job Corps' reported graduates were placed into jobs, went on to further education, advanced training, or military service. The Home Builders Institute (HBI) says that those completing training earn fifty-two percent more than those who don't finish training.

The National Association of Home Builders is the vocational trainer in Job Corps. NAHB's education arm is the HBI. The Job

Corps is HBI's largest training program. It has prepared more than fifty thousand young people for careers in the building industry. HBI's craft skills program is its oldest construction trades training program. These pre-apprenticeship and nonunion apprenticeship training programs are operated by more than one hundred builder associations. For more information from the Home Builders Institute: 15th and M Streets, NW, Washington, D.C. 20005 (202-822-0494 or 800-368-5242, ext. 494).

In March 1988, Fred Day, the national coordinator for the Home Builders Institute, said, "Out of eleven trades that we train in, masonry is the number one. We've got about two jobs for every trainee. Demand is biggest in the Northeast and North. Wages usually are five dollars to eleven dollars an hour."

In 1988 the Job Corps had 174 trainees in seven training centers: Roswell, New Mexico; Jacksonville, Florida; Old Dominion, Virginia; Washington, D.C.; Scranton, Pennsylvania; and two in Baltimore. Each trainee spends a minimum of eight hundred hours on the job with little classroom work. Now there are many more training centers for brick and cement masonry pre-apprenticeship and apprenticeships. See appendix D.

Most Job Corps graduates were high school dropouts eighteen years of age, who read at the elementary school level, had never held a full-time job, and came from families of four or more with an income below five thousand dollars a year. About ninety-five percent are high school dropouts and a third come from families on public assistance. More than seventy percent are minorities and about thirty percent are women.

The International Masonry Institute was awarded $4.1 million in July 1990 to operate forty-two craft training programs at thirty-four Job Corps Centers. Of these programs, thirty-five are for bricklaying. These forty-two programs provided 840 training slots for 1991. In the preceding contract term, 1,186 people enrolled in IMI Job Corps programs and 475 were placed with an average apprenticeship starting wage of $9.86.

HOW DO I ENTER THE JOB CORPS?

Call the Job Corps for more information by calling this toll-free number: 1-800-TRAIN YOU (1-800-872-4696). The person answering the call can explain eligibility requirements and benefits and refer you to where you can get an interview to see if you qualify. This screening interview ensures a good match between the applicant and the Job Corps center.

The Job Corps also recruits people through referrals from public and private agencies. You can also apply directly by writing to the Job Corps, Employment and Training Administration, U.S. Department of Labor, Washington, D.C. 20213.

To be eligible for entry into the Job Corps program you must meet these qualifications:

- must be 16 and not yet 22 years of age at time of enrollment.
- be a United States citizen, United States national, permanent resident alien, or other alien who can accept permanent employment in the United States.
- be economically disadvantaged.
- want and be capable of acquiring additional skills training to meet entrance requirements for the military or qualify for a job that requires education or skill training.
- have signed consent from a parent or guardian if applicant is under eighteen.
- not be on probation, parole, under a suspended sentence, or under the supervision of any court agency or institution (unless special permission is granted).
- live in an environment that is not conducive to getting an education or a job.

WHAT WILL THE JOB CORPS DO FOR ME?

Many programs are available through the Job Corps, including brick masonry, cement masonry, and plastering. Since the program is government funded, you have no expenses. This is what the program provides for you:

- an individualized self-paced curriculum with instructors leading to a GED or advanced educational training
- free room and meals
- up to one hundred dollars spending money each month (based on an incentive system)
- up to two thousand four hundred dollars allowance upon graduation (depending on length of stay) to be used for a car down payment, security deposit on a new apartment, or other expense incurred while adjusting to working life
- about three hundred dollars clothing allowance for the first year in Job Corps
- money sent home for child care
- driver's education
- free transport to and from the Job Corps center
- free books, work clothing, and tools

An added advantage to the Job Corps program is that it helps improve your reading, writing, and math skills. Having these skills helps you get ahead in any job you choose. This is especially important to Job Corps members, many of whom start with poor reading, writing, and math skills.

When you complete the program, you can qualify for a union apprentice program or you may work for a contractor not requiring union workers. You will not be at the skill level of a union journeyman mason, but you will have many of the skills needed to build many masonry projects.

WHERE DO I LIVE WHILE IN THE JOB CORPS?

You can be part of the Job Corps on a residential or nonresidential basis. In the 1960s, when the Job Corps first took off, the residences were in the cities where most of the students came from. Then these training centers were moved to rural and wilderness areas because it was found that getting the students further away from a bad environment helped them stay in the program and get more out of it.

On a residential basis, you live in a dorm room with roommates. The dorms generally have laundry facilities, lounges, telephone areas, and bathrooms. Meals are served three times a day. Typical Job Corps rules include: no alcoholic beverages or drugs at the center; no weapons; and no fighting, gambling, stealing, or sexual activity.

Most centers have a dress code during working hours that reinforces the proper dress expected on the job.

But this is not like prison. Corpsmembers have an active voice in recommending changes in the operation of the center through the corpsmember government. This group, composed of student-elected peers, also sponsors dances, picnics, and field trips.

WHAT DO JOB CORPS CLASSES TEACH ME?

The basic education course covers cultural awareness (getting along), health education (nutrition, first aid, and family roles), world of work (safety, interviewing techniques, taxes), and driver education. All centers do not offer advanced training opportunities, and advanced training is not available in all trades.

In the masonry skills training, you are judged as having limited skill, moderate skill, or skill in the following outline of activities:

I. Mason Tending
   A. demonstrate correct use of hand tools
   B. mix mortars to proper proportions
   C. prepare job to start work
   D. operate equipment safely and correctly
   E. set up and break down scaffolding

    F. clean up job
    G. inspect scaffolding for safety problems
    H. stock job to keep it running smoothly
II. Masonry Measuring and Cutting
    A. list safety rules related to masonry cutting
    B. operate masonry saw
    C. cut masonry with hammer and brick set
    D. demonstrate toothing masonry
    E. remove mortar joints
    F. use and read a foot rule correctly
    G. read a modular and spacing rule correctly
III. Brick, Block, and Stone
    A. spread mortar
    B. butter masonry units
    C. lay units to a line
    D. tool joints
    E. rake out joints
IV. Installation and Building
    A. describe safety precautions associated with building and installing masonry units
    B. lay up veneer
    C. lay overhand
    D. lay a rowlock
    E. install sills
    F. install headers
    G. install underpinning
    H. build corners and leads
    I. build pilasters and piers
V. Layout and Bond
    A. lay out a brick wall
    B. lay out a block wall

      C. lay out a composite wall
      D. lay an American bond
      E. lay an English bond
      F. lay a Flemish bond
      G. lay a pattern bond
      H. lay out and install pavers using a paver rack in various patterns

VI. Reinforcement
      A. describe the safety precaustions associated with masonry reinforcement
      B. install lateral wall ties
      C. install horizontal wall reinforcement
      D. install rebar
      E. install lintels
      F. install bond beams

VII. Measuring and Leveling
      A. make and use a story pole
      B. establish building lines
      C. use levels and transits
      D. demonstrate 6-8-10 squaring method

VIII. Pointing, Cleaning, Caulking
      A. describe safety precautions related with pointing, cleaning, and caulking
      B. use acids and detergents for cleaning
      C. demonstrate sandblasting for cleaning
      D. demonstrate mortar matching
      E. demonstrate masonry replacement

IX. Grouting
      A. mix reinforced grout
      B. mix grouts for structural tile
      C. mix grouts for brick pavers

D. grout reinforced masonry
  E. grout structural tile
  F. grout brick pavers
X. Fireplaces (Optional)
  A. describe safety precautions associated with masonry fireplaces
  B. lay out a firebox
  C. install a firebox
  D. set dampers and headforms
  E. construct a brick corbel
  F. place a flue liner

All of these skills are also taught in the union's three-year apprenticeship program. The Job Corps' program does not last as long so you do not get as long a chance to practice. That is why this program is considered a pre-apprentice program. You are introduced to all the elements of building with masonry, but you have yet to perfect them.

In addition to masonry skills and basic education skills, you will also be judged on your employability skills. You are judged on the following characteristics as: never, sometimes, usually, and always.

I. Grooming and Hygiene
  A. maintains good grooming and personal hygiene
  B. dresses properly for the trade
II. Dependability
  A. arrives on time
  B. remains at work throughout duty time
  C. completes assigned tasks
  D. follows instructons
  E. follows safety rules

III. Adaptability
   A. changes from one task to another easily
   B. adapts to a variety of job tasks and situations
IV. Interpersonal Skills
   A. shows positive attitude toward work and people
   B. works cooperatively with fellow students and instructor
   C. listens and asks questions
   D. treats others with respect
   E. accepts constructive criticism
   F. respects worth of equipment and materials

WHAT HAPPENS AFTER I GRADUATE FROM JOB CORPS?

After graduation, all Job Corps members are given assistance to find a job and start working life (such as developing budgets, purchasing insurance, and finding an apartment).

In operation since 1964, Job Corps is the only federally funded job training and education program with an active alumni association—more than twenty-six thousand alumni members in fifty chapters nationwide.

CHAPTER THIRTEEN

EARNING FAME AND OTHER CERTIFICATIONS

CONTESTS FOR SKILL AND DESIGN

In addition to having your journeyman mason's card or a certificate of completion of a masonry program, you can earn satisfaction by entering various masonry industry contests. Some are local and some are national. Some have to do with skills, many have to do with design.

The design awards are usually granted to the architect, but the contracting firm is often listed prominently. There are state, regional, and national competitions for outstanding examples of masonry construction.

You do not have to be a journeyman mason to enter contests. Vocational Industrial Clubs of America (VICA) holds an annual VICA United States Skills Olympics. The Brick Institute of America and National Concrete Masonry Association also cosponsor the event.

In 1989 VICA introduced a new event: composite wall construction. In the brick and block competition, students had to follow specifications on drawing to build a wall system that used about thirty block and two hundred and fifty brick. These students plan, lay out, and build the structure and then clean up their work. Contestants are judged on height, level, plumb, neatness, correct design, square and range, manipulation, uniform joints, and production. They are even evaluated on use of proper work clothes and safety techniques. A written test is also part of the evaluation.

Another contest of skills is the "Fastest Trowel on the Block" competition held during the fall of each year. In 1992 it was held September 11–12 in Spartansburg, South Carolina. It is sponsored by the Masonry Construction Association of America (MCAA).

Masonry has even made it into the *Guiness Book of World Records*. The book features Bob Boll, Geoffrey Capes, David and Kym Barger, and Cynthia Ann Smolko for brick-related achievements.

Boll laid 914 bricks in one hour at the National Speed Bricklaying contest in Lansing, Michigan. The previous record was 725 bricks. Capes threw a five-pound brick 146 feet, one inch on July 19, 1978. The Bargers carried an eight-pound, fifteen-ounce brick in a nominated, ungloved hand extended in an uncradled, downward pincer grip for forty-five miles. Smolko used the same grip to carry a nine-pound, twelve-ounce brick 19.2 miles.

CERTIFICATION PROGRAMS

You can also earn other certifications relevant to your work. These certifications expand your knowledge and skills. They also serve as testament to your clients (if you are a contractor) or to your employer (if you work for a contractor) that you have gone that extra step in advancing yourself and deserve more of their business or a better position.

Oftentimes, local masonry associations have programs that let contractors earn certification in various specialty areas and this certification can earn you more work. A certification from an association or governmental body indicates to your clients—other contractors and owners—that you do quality work.

For example, in St. Louis masonry contractors can become certified by the Masonry Contractor Association. To become a certified mason contractor in St. Louis the contractors must:

- be active in masonry construction and contracting in the St. Louis area for at least two years
- show proof of eligibility
- earn a passing grade on a comprehensive written test
- renew certification every two years

The test takes about one-and-a-half hours and has 125 questions, but twenty-five of them can be ignored to make the test equitable for everyone because one masonry contractor may specialize in cleaning, tuckpointing, and preservation and another may do granite high-rises. The test covers:

- administrative planning and supervision
- fiscal and financial responsibilities
- planning and supervising construction
- bidding, contract negotiations, and ethical standards
- safety practices and procedures
- masonry materials and applications
- building code requirements

Passing grade is seventy percent. To recertify, the contractor must earn twenty recertification credits in the following two years by attending approved educational programs, seminars, and workshops.

The program commits the contractor to self-improvement. Masonry does change, contrary to popular belief, and new developments can only be learned by attending educational programs regularly.

This certification program promotes increased public confidence in the managerial and technical skills of masonry contractors. Currently being certified is just a bonus that contractors can use to sell their services. However, some job specifications are beginning to require that the masonry subcontractor on the job be certified.

For more information about MCA's Masonry Contractor Certification Program, write MCA, 1429 S. Big Bend, St. Louis, MO 63117 (314-645-1966).

APPENDIX A

GLOSSARY

Admixture. An ingredient (liquid or powder) added to mortar to change its properities. Some admixtures extend the time before the mortar sets, others speed it up. Some add water repellency, make the mortar more workable, or color the mortar.

Aggregate. The inert stone or sand added to mortar or a concrete mix.

Apprentice. Person contracted to a training program run by a Joint Apprenticeship and Training Committee (JATC) in the building trades. See *journeyman*.

Ashlar. A square or rectangular cut stone used in building.

Bat. A segment of a brick. Often half of a brick.

Batter. Sloping back of the outer side of a masonry wall.

Bed. The bottom side of a brick or block that is laid in mortar in the wall.

Benefits. These are the extras you get from your employer in addition to your wages. It can include medical, dental, vision, unemployment, and life insurance, pension, short- and long-term disability pay, and vacation. Not all employers provide the same coverage, if they supply any at all. If you are a union journeyman mason, your union will supply you with many of these benefits.

Block. Concrete block. Comes in various colors and textures: smooth, ground, split face, ribbed, fluted, and adobe "slump."

Blue-collar worker. A person employed in a production, construction, or maintenance position. The opposite would be a "white-collar" employee, who would be an office worker or professional person. Women in lower-paying white-collar work, such as secretaries, are sometimes called "pink-collar" workers.

Blueprint. A reproduction of an architect's plans of a structure. The photographic process by which they are made produce white lines on a blue background.

Bond. Pattern of laid masonry units. Various bonds include common, English, Flemish, Dutch, garden wall, and monk. It also refers to the adhesion between mortar and masonry units.

Brick. A fired clay rectangularly shaped masonry unit, with or without cores. Comes in more than ten thousand combinations of shape, size, color, and texture. Three main types: building, face, and paving. Various colors ranging from white, to earth hues, to black and exotic glazed shades; textures and shapes; laid in various patterns.

Brick tongs. A device that holds many brick; like a clamp.

Butter. Used as a verb, butter means to spread mortar on to a brick with a trowel before setting it in place, "butter the brick."

Cavity wall. The space between two wythes of masonry. Usually a two-inch space. Also called a hollow wall.

Cement. A very fine powder that, when mixed with lime, water and sand, becomes mortar. Mixed with water and larger aggregate, it becomes concrete. Usually called portland cement.

Cement masonry. A construction trade that builds with concrete.

Chase. Set back built into wall for piping, electric wires, and ducts.

Clinker brick. Overfired brick, on purpose or by accident, that become fused, twisted, or glazed. Both decorative, and used carefully and structurally.

Closure brick. The last brick or block laid in a course.

CMU. Abbreviation for concrete masonry unit; concrete block.

Collective agreement. A contract between a union and an employer or group of employers that sets down employee wages, hours, working conditions, and benefits.

Composite wall. Masonry wall with wythes of different materials, such as brick and block. Rigid foam insulation board can be inserted in the two-inch cavity, leaving a one-inch cavity.

Concrete. A building material that combines cement, water, sand and aggregate. It can be shaped on site in forms or precast in a manufacturing plant.

Contractor. A person or company who agrees, by contract (usually written), to do certain work under specified conditions and prices by a date.

Coping. Masonry cap on the top of a wall or pier.

Corbel. A ledge built out from the face of a masonry wall that projects out further and further as they go higher; the opposite of *batter*.

Corner pole. A measuring stick that helps the mason keep course height of bricks consistent.

Course. Horizontal row of masonry units.

Craft. An occupation or trade requiring a higher level of skill or artistic ability.

Darby. A trowel that compacts and levels poured concrete.

Efflorescence. Stain that forms on mortar or concrete; caused by moisture leaching salts from the masonry or mortar. This stain is usually white colored, but can be greenish too.

Face. The exposed surface of a masonry unit or the opening of a fireplace.

Fireback. The back wall of the firebox.

Firebox. The chamber where the fire is built.

Firebrick. Brick made of fire-resistant clay that lines the firebox in a fireplace.

Flashing. Installed anywhere in the wall wherever there are interruptions such as sills, shelf angles, spandrels, parapet walls, and lintels to keep water out of the walls. Material can be zinc coated, copper coated with polyethylene so as not to stain brick, a plastic, or a combination of materials. Stainless steel flashing costs a lot, but lasts longest.

Float. A board used by cement masons that smooths the surface of a concrete slab. A big one is called a bull float. It can be attached to a pole so that the cement mason can stand and smooth large surfaces of recently placed concrete.

Flue lining. The channel inside the chimney that carries smoke and gases to the outside. Usually made of 5/8-inch-thick clay liners.

Footing. Support for walls, columns, or piers. Often made of concrete or concrete block.

Foreman. Experienced journeyman who directs the work crew and indicates what work is to be done each day. The foreman reports to the superintendent. Foremen usually earn seventy-five cents to two dollars more per hour than journeymen.

Frog. Small indentation in the bed of a brick.

Green. Describes bricks that have not been fired yet and mortar that has not set.

Grout. Mortar that can be poured to fill masonry voids.

Header. Masonry unit laid with end facing out.

Hod. V-shaped, often long-handled, carrier for mortar. Used to carry the mixed mortar up to where the mason is working. A long handle is attached to the hod so it can be lifted up to the work area.

Hydration. The chemical reaction that takes place between water and cement. When the cement has reacted with all the water the reaction is complete and the mortar or concrete is hardened.

Insulation. Block inserts or polystyrene sheets between wythes.

Joint. The mortar-filled space between the bricks. An expansion joint can be a horizontal or vertical joint that runs the length or width of a wall and allows for movement.

Jointing. Finishing of masonry joints. Also called tooling. Joints can be flushed, weathered, concave, V, struck, raked, or extruded (or weeping).

Journeyman. Craftsman who has completed and passed an apprenticeship in a trade. Journeymen are entitled to the highest minimum wage rate established for their job classification. All journeymen working on a job site are paid the same rate, no matter how long they have been a journeyman. The union negotiates their hourly rate with the area contractors.

Lead. A built-up masonry corner that is used as a guide in laying a wall. (Pronounced "leed.")

Level. Exactly horizontal.

Line blocks. Fit around a corner or jamb and tie the line to leads or corner poles.

Lintel. A steel angle or reinforced masonry beam located above a fireplace, door, or window opening to support it.

Mason. A skilled worker, usually a journeyman, who works with and lays brick, concrete block, glass block, stone, or other masonry units. A cement mason places and finishes concrete.

Masonry. Craft of laying masonry units, such as brick, concrete block, glass block, structural tile, and stone. The word also refers to the building material itself.

Mortar. A combination of sand, cement, and water used to bond masonry units together. It can also have additives in it to change its color and properties. There are four types of mortar currently, types M, S, N, and O. They are named after the words MaSoN wOrK (K has been eliminated). They are used for different purposes. They are not numbered or alphabetized so that no one ranks them in order of importance—each type has its own purpose.

Mortar board. Square board used by masons to hold mortar ready for use on the job.

Mud. Wet concrete or wet mortar.

Parapet. Wall section that extends above the roof; usually seen on flat roofs.

Paver. Special masonry units used for floors, walkways, and patios. They are made of precast concrete or clay brick. Brick pavers may have mortar in the joints or just swept sand.

Pier. A short masonry or concrete column supporting the foundation of the floor structure in spaces without a basement.

Pilaster. A pier or column forming part of a masonry or concrete wall. It projects out from it and is designed to receive a joist or beam load.

Plumb. Exactly vertical.

Pointing. To repair mortar joints by removing crumbled mortar and filling the gaps with new mortar.

Pre-apprentice. Nonunion training that prepares a person for apprenticeship training. Not a requirement before entering an apprenticeship, but it prepares people not otherwise qualified.

Quoin. Large squared stone or brick set at the corner formed by two masonry walls. Often projects from the corner slightly.

Rebar. A metal reinforcing rod used to add shear and tension strength to a masonry wall or concrete.

Retemper. To add more water to mortar to make it more workable.

Rowlock. A header turned on its side.

Sailor. Soldier with wide side forward. (See *soldier.*)

Scab. A person who accepts employment or replaces a union worker during a strike. A derogatory term used by union workers. Scabs work for less than union wages or on nonunion terms.

Scaffolding. A temporary or movable platform for workers to stand on while working on a building's wall. Some can be raised or lowered on a winch; others need to have the flooring platforms moved manually higher. Scaffolds have railings to protect workers from falling off.

Set. Occurs when mortar or concrete has fully hydrated (often called hardened).

Smoke shelf. Prevents a downdraft from entering firebox and blowing smoke into the room. May be curved or flat.

Soldier. A brick that stands upright rather than on its side. If it is not exactly straight up and down, it is called a drunken soldier.

Specifications. Written description of work to be done, quality of work, and type of materials. Often called specs.

Stone. Finishes may be polished, flamed, bush-hammered, hand-tooled, or chiseled.

Stretcher. A standard size brick laid on its side.

Subcontractor. A person or company who contracts with a general contractor to do a certain portion of work for a set price. Plumbers, masons, and electricians are typically subcontractors. Subcontractors usually do specialty work, such as tuckpointing.

Superintendent. Supervises the work of many sites or many crafts on one site for the contractor. The superintendent is usually directed by the contractor, unless the contracting firm is large.

Tender. Laborer who helps a mason. They ensure materials are available to the mason and do odd jobs around the job site like cleaning up.

Terra-cotta. Fired clay building unit that can be molded into any shape before fired. Often ornamental.

Tile. Hollow masonry unit made of fired clay, shale, fire clay, or a mixture of these.

Trig. Device to support a line at the center of a wall.

Trig unit. Masonry units set in the middle of the wall to hold up the line. Prevents line from sagging and prevents wind from blowing the line. The metal clip it is attached to is called a twig. Used for lines longer than forty feet.

Tuckpointing. Refilling masonry joints that have been raked out and cleaned. *Repointing* is the term for both raking out the joints and refilling them.

Veneer. Separate wythe of masonry units that is not load-bearing; used as a facing for buildings and attached to a structural backing such as wood or steel.

Wage scale. A listing of wages that shows the pay (usually hourly rates) for particular job classifications.

Weep holes. Holes in the masonry joints that allow water to escape a wall.

Wythe. Vertical wall of masonry units that are one-unit thick.

APPENDIX B

BRICKLAYER UNION BUREAUS OF APPRENTICESHIP AND TRAINING

Region 1
(Connecticut, Maine, Massachusetts, New Hampshire, Rhode Island, Vermont)
United States Department of Labor
J.F. Kennedy Building, Room 510
Government Center
Boston, MA 02203
617-565-2288

Region 2
(New Jersey, New York, Puerto Rico, Virgin Islands)
United States Department of Labor
201 Varick Street, Room 602
New York, NY 10014
212-337-2313

Region 3
(Delaware, Maryland, Pennsylvania, Virginia, West Virginia, District of Columbia)
United States Department of Labor
Gateway Bldg.
3535 Market Street, Room 13240
Philadelphia, PA 19104
215-596-6417

Region 4
  (**Alabama, Florida, Georgia, Kentucky, Mississippi, North Carolina, South Carolina, Tennessee**)
  United States Department of Labor
  1371 Peachtree Street, NE, Room 418
  Atlanta, GA 30367
  404-347-4405

Region 5
  (**Illinois, Indiana, Michigan, Minnesota, Ohio, Wisconsin**)
  United States Department of Labor
  Federal Building, Room 758
  230 South Dearborn Street
  Chicago, IL 60604
  312-353-7205

Region 6
  (**Arkansas, Louisiana, New Mexico, Oklahoma, Texas**)
  United States Department of Labor
  Federal Building, Room 502
  525 Griffin Street
  Dallas, TX 75202
  214-767-4993

Region 7
  (**Iowa, Kansas, Missouri, Nebraska**)
  United States Department of Labor
  Federal Office Building, Room 1100
  911 Walnut Street
  Kansas City, MO 64106
  816-374-3856

Region 8
  (**Colorado, Montana, North Dakota, South Dakota, Utah, Wyoming**)
  United States Custom House
  Room 476
  721 19th Street
  Denver, CO 80202
  303-844-4791

Region 9
(**Arizona, California, Hawaii, Nevada, Guam**)
United States Department of Labor
71 Stevenson Street, Room 715
San Francisco, CA 94105
415-995-5542

Region 10
(**Alaska, Idaho, Oregon, Washington**)
United States Department of Labor
909 First Avenue, Room 8018
Seattle, WA 98174
206-442-5286

STATE OFFICES

Alabama
Berry Building
2017 2nd Ave., North,
 Suite 102
Birmingham, AL 35203
205-731-1308

Alaska
Federal Building and
 Courthouse, Box 37
Room C-528
701 C Street
Anchorage, AK 99513
907-271-5035

Arizona
3221 North 16th Street, Suite 302
Phoenix, AZ 85016
602-241-2964

Arkansas
Federal Building
Room 3309
700 West Capitol Street
Little Rock, AR 72201
501-378-5415

California
211 Main Street,
 Room 350
San Francisco, CA 94105
415-974-0556

Colorado
U.S. Custom House
721 19th Street,
 Room 480
Denver, CO 80202
303-844-4793

Connecticut
Federal Building
135 High Street,
 Room 367
Hartford, CT 06103
203-240-4311

Delaware
Federal Building
844 King Street,
 Lock Box 36
Wilmington, DE 19801
302-573-6113

Florida
City Centre Building
227 North Bronough Street,
  Room 1049
Tallahassee, FL 32301
904-681-7161

Georgia
1371 Peachtree Street, NE,
  Room 418
Atlanta, GA 30367
404-881-4403

Hawaii
P.O. Box 50203
300 Ala Moana Boulevard,
  Room 5113
Honolulu, HI 96850
808-541-2518

Idaho
P.O. Box 006
550 West Fort Street, Room 493
Boise, ID 83724
208-334-1013

Illinois
230 S. Dearborn Street,
  Room 702
Chicago, IL 60604
312-353-4690
312-353-7205
312-353-1572

Indiana
Federal Building and U.S.
  Courthouse
46 East Ohio Street, Room 414
Indianapolis, IN 46204
317-269-7592

Iowa
Federal Building
210 Walnut Street, Room 637
Des Moines, IA 50309
515-284-4690

Kansas
Federal Building
444 S.E. Quincy Street,
  Room 235
Topeka, KS 66683
913-295-2624 (Ext. 236)

Kentucky
Federal Building
600 Federal Place,
  Room 187-J
Louisville, KY 40202
502-582-5223

Louisiana
U.S. Postal Building
701 Loyola Street, Room 1323
New Orleans, LA 70113
504-589-6103

Maine
Federal Building
P.O. Box 917
68 Sewall Street, Room 408-D
Augusta, ME 04330
207-622-8235

Maryland
Charles Center Federal Building
31 Hopkins Plaza, Room 1028
Baltimore, MD 21201
301-962-2676

Massachusetts
JFK Federal Building
Government Center,
  Room 1703-B
Boston, MA 02203
617-565-2291

Michigan
Federal Building
231 W. Lafayette Avenue,
  Room 657
Detroit, MI 48226
313-226-6206

Minnesota
Federal Building and U.S. Courthouse
316 Robert Street, Room 134
St. Paul, MN 55101
612-290-3951

Mississippi
Federal Building
100 West Capitol Street, Suite 1010
Jackson, MS 39269
601-960-4346

Missouri
210 North Tucker, Room 547
St. Louis, MO 63101
314-425-4522

Montana
Federal Office Building
301 South Park Avenue, Room 394 Drawer #10055
Helena, MT 59626-0055
406-449-5261

Nebraska
106 South 15th Street, Room 700
Omaha, NE 68102
402-221-3281

Nevada
Post Office Building
P.O. Box 1987
301 East Stewart Avenue, Room 311
Las Vegas, NV 89101
702-388-6396

New Hampshire
Federal Building
55 Pleasant Street, Room 311
Concord, NH 03301
603-225-1444

New Jersey
Military Park Building
60 Park Place, Room 339
Newark, NJ 07102
201-645-3880

New Mexico
320 Central Avenue, SW, Suite 16
Albuquerque, NM 87102
505-766-2398

New York
Federal Building
North Pearl & Clinton Avenues, Room 810
Albany, NY 12201
518-472-4800

North Carolina
Federal Building
310 New Bern Avenue, Room 376
Raleigh, NC 27601
919-856-4466

North Dakota
New Federal Building
653 2nd Avenue, North, Room 344
Fargo, ND 58102
701-239-5415

Ohio
200 North High Street, Room 605
Columbus, OH 43215
614-469-7375

Oklahoma
Alfred P. Murrah Federal Building
200 N.W. 5th Street, Room 526
Oklahoma City, OK 73102
405-231-4814

Oregon
Federal Building
1220 SW 3rd Avenue, Room 526
Portland, OR 97204
503-221-3157

Pennsylvania
Federal Building
228 Walnut Street, Room 773
Harrisburg, PA 17108
717-782-3496

Rhode Island
Federal Building
100 Hartford Avenue
Providence, RI 02909
401-273-7640

South Carolina
Strom Thurmond Federal
Building
1835 Assembly Street,
Room 838
Columbia, SC 29201
803-765-5547

South Dakota
Courthouse Plaza
300 North Dakota Avenue,
Room 403
Sioux Falls, SD 57102
605-330-4326

Tennessee
460 Metroplex Drive, Room 606
Nashville, TN 37203
615-736-5408

Texas
VA Building
2320 LaBranch Street,
Room 2102
Houston, TX 77004
713-750-1696

Utah
1745 West 1700 South,
Room 1051
Salt Lake City, UT 84104
801-524-5700

Vermont
Burlington Square
96 College Street, Suite 103
Burlington, VT 05401
802-951-6278

Virginia
400 North 8th Street,
Room 10-020
Richmond, VA 23240
804-771-2488

Washington
Federal Office Building
909 First Avenue,
Room B-104
Seattle, WA 98174
206-442-4756

West Virginia
550 Eagan Street, Room 310
Charleston, WV 25301
304-347-5141

Wisconsin
Federal Center
212 East Washington Avenue,
Room 303
Madison, WI 53703
608-264-5377

Wyoming
J.C. O'Mahoney Federal Center
P.O. Box 1126
2120 Capitol Avenue,
Room 8017
Cheyenne, WY 82001
307-772-2448

APPENDIX C

STATE VICA DIRECTORS

If you call, ask for the state VICA director or send your letter to "State VICA Director." VICA programs provide you with pre-apprenticeship training. See chapter eleven for more information.

Alabama
State Department of Education
Room 5226, Gordon Persons
  Building
Montgomery, AL 36130
205-242-9112

Alaska
1201 West Vaunda
Wasilla, AK 99687
907-376-5341 or 907-376-3563

Leadership Experience, Inc.
520 East 34th Avenue, #107
Anchorage, AK 99503
907-563-7882

Arizona
Department of Education/
  Vocational Education
1535 West Jefferson
Phoenix, AZ 85007
602-542-5422

Arkansas
Three State Capitol Mall
Little Rock, AR 72201-1083
501-682-1271

California
721 Capitol Mall, 4th Floor
Sacramento, CA 95814
916-657-2575

Canada
Skills Canada
Central Technical School
725 Bathurst Street
Toronto, Ontario, Canada M5S 2R5
705-393-0125

Colorado
Park Center Building,
  Suite 600
1391 Speer Boulevard
Denver, CO 80204
303-620-4000

Connecticut
25 Industrial Park Road
Middletown, CT 06457
203-638-4057

Delaware
Department of Public
 Information
The Townsend Building
Loockerman Street
P.O. Box 1402
Dover, DE 19903
302-739-4681

District of Columbia
Taft Junior High School
18th & Perry streets, NE
Washington, D.C. 20018
202-576-6278

Florida
State Department of Education
Turlington Educational Center
325 West Gaines Street,
 Suite 1232
Tallahassee, FL 32399
904-488-2677

Georgia
State Department of Education
Division of Vocational Education
1752 Twin Towers, East
Athens, GA 30334
404-656-2554

State Department of Technical
 and Adult Education
660 South Tower
One CNN Center
Atlanta, GA 30303
404-656-2550

Hawaii
Department of Education
49 Funchal Street, #J-305
Honolulu, HI 96813
808-586-3563

Idaho
2308 Terrace Drive
Caldwell, ID 83605
208-459-2418

Illinois
Don Bauc
3612 Chestnut Drive
Hazel Crest, IL 60429
708-335-1687

Indiana
Division of Vocational
 Education
State House, Room 229
Indianapolis, IN 46204
317-232-9184

Iowa
Lynette Gibb
P.O. Box 1871
Ames, IA 50010
515-292-8467

Kansas
204-205 Willard Hall
Pittsburg State University
Pittsburg, KS 66762
316-235-4636

Kentucky
2625 Capital Plaza Tower
Frankfort, KY 40601
502-564-2890

Louisiana
626 North 4th Street,
 Room 300
Baton Rouge, LA 70802
504-342-1499

State Department of Education
Postsecondary Vocational
 Education
P.O. Box 94064
Baton Rouge, LA 70804
504-342-3544

State VICA Directors

Maine
Bureau of Vocational
  Education
Department of Education
State House Station, #23
Augusta, ME 04333
207-289-5854

Southern Maine Technical
  College
Fort Road
South Portland, ME 04106
207-799-7303

Maryland
MSDA-OVTE
200 West Baltimore Street
Baltimore, MD 21201
301-333-2572

Massachusetts
Department of Occupational
  Education
1385 Hancock Street
Quincy Center Plaza
Quincy, MA 02169
617-770-7366

Michigan
1115 South Pennsylvania
  Avenue, Suite C
Lansing, MI 48912
517-487-5893

Minnesota
Capitol Square Building
550 Cedar Street, 556A
St. Paul, MN 55101
612-296-1067

Mississippi
Division of Vocational
  Education
550 High Street, Suite 1001
P.O. Box 771
Jackson, MS 39205
601-359-3480

Missouri
State Department of Education
205 Jefferson
P.O. Box 480
Jefferson City, MI 65102
314-751-4460

Montana
Office of Public Instruction
State Capitol
Helena, MT 59620
406-444-4452

Nebraska
301 Centennial Mall South
P.O. Box 94987
Lincoln, NE 68509
402-471-4820

Nevada
State Department of Education
400 West King Street
Carson City, NV 89710

New Hampshire
Dover High School
Durham Road
Dover, NH 03820
603-742-3176, ext. 327

New Jersey
New Jersey Department of
  Education
Division of Vocational Education
225 West State Street, CN500
Trenton, NJ 08625
609-292-6594

New Mexico
820 Manzano, NE
Albuquerque, NM 87110
505-265-9225

New York
Joan Grimes
RD 2, Box 514
Cooperstown, NY 13376
607-547-9700

North Carolina
116 West Edenton Street
Education Building, Room 580
Raleigh, NC 27603
919-733-7421

North Dakota
State Board for Vocational
    Education
Capitol Building, 15th Floor
600 East Boulevard Avenue
Bismarck, ND 58505
701-224-3163

Ohio
65 South Front Street, Room 915
Columbus, OH 43266
614-466-2901

Oklahoma
1500 West 7th Avenue
Stillwater, OK 74074
405-743-5143

Oregon
700 Pringle Parkway, SE
Salem, OR 97301
503-378-5114

Pennsylvania
State Department of Education
333 Market Street, 6th Floor
Harrisburg, PA 17126
717-787-8804

Puerto Rico
Juan A. Pineiro Gonzalez
Box 759
Hato Rey, PR 00919
809-754-1270

Rhode Island
State Department of Education
22 Hayes Street, Room 222
Providence, RI 02908
401-277-2705

South Carolina
State Department of Education
1831 Barnwell Street
Columbia, SC 29201
803-253-4025

South Dakota
Division of Vocational
    Education
Richard F. Kneip Building
700 Governors Drive
Pierre, SD 57501
605-773-3423

Tennessee
State Department of Education
531 Henley Street
Knoxville, TN 37902
605-594-6044

Texas
Texas Education Agency
1701 North Congress, #3-110
Austin, TX 78701
512-463-9688

Utah
Scott Nielson
45 East State Street
Farmington, UT 84025
801-451-1291

Vermont
Oxbow Vocational Center
P.O. Box 618
Bradford, VT 05033
802-222-5212

Virgin Islands
Vocational Technical
    Education
V.I. Department of Education
Christiansted
St. Croix, VI 00820
809-773-0500

Virginia
State Department of
  Education
101 North 14th Street,
  19th Floor
Richmond, VA 23219
804-225-2161

Washington
New Market Skills Center
7299 New Market Street
Turnwater, WA 98501
206-586-9373

West Virginia
Bureau of Vo-Tech & Adult
  Education
Capitol Complex, Building 6
1900 Kanawha Street, B-243
Charleston, WV 25305
304-348-6313

Wisconsin
Department of Public Instruction
125 South Webster Street
P.O. Box 7841
Madison, WI 53707
608-267-9251

APPENDIX D

JOB CORPS OFFICES

To join a Job Corps program you must choose one in your region. You do not have to stay within your state. There are strict requirements to be eligible for the Job Corps training programs. For more information, call 1-800-TRAIN-YOU (1-800-872-4696) and see Chapter Twelve.

The state offices are grouped by the programs they provide: bricklaying pre-apprenticeship, bricklaying apprenticeship, cement masonry pre-apprenticeship, and cement masonry apprenticeship.

The pre-apprentice programs are for young people with no experience. The apprentice programs are for those young people with experience and you earn your journeyman card, allowing you to work as a union mason.

Region 1
  (Connecticut, Maine, Massachusetts, New Hampshire, Rhode Island, Vermont)
United States Department of Labor, ETA
Office of Job Corps
J.F. Kennedy Building,
  Room 1700-C
Government Center
Boston, MA 02203
617-565-2166

Region 2
(**New Jersey, New York, Puerto Rico, Virgin Islands**)
United States Department of Labor, ETA
Office of Job Corps
201 Varick Street, Room 897
New York, NY 10014
212-337-2282

Region 3
(**Delaware, Maryland, Pennsylvania, Virginia, West Virginia, District of Columbia**)
United States Department of Labor, ETA
Office of Job Corps
3535 Market Street, Room 12220
Philadelpha, PA 19104
215-596-6301

Region 4
(**Alabama, Florida, Georgia, Kentucky, Mississippi, North Carolina, South Carolina, Tennessee**)
United States Department of Labor, ETA
Office of Job Corps
1371 Peachtree Street, NE, Room 632
Atlanta, GA 30309
404-347-3178

Region 5
(**Illinois, Indiana, Michigan, Minnesota, Ohio, Wisconsin**)
United States Department of Labor, ETA
Office of Job Corps
Federal Building, Room 676
230 South Dearborn Street
Chicago, IL 60604
312-353-1572

Region 6
(**Arkansas, Louisiana, New Mexico, Oklahoma, Texas**)
United States Department of Labor, ETA
Office of Job Corps
Federal Building, Room 317
525 Griffin Street
Dallas, TX 75202
214-767-2567

Region 7
   (Iowa, Kansas, Missouri, Nebraska)
   United States Department of Labor, ETA
   Office of Job Corps
   Federal Office Building, Room 1102
   911 Walnut Street
   Kansas City, MO 64106
   816-426-3661

Region 8
   (Colorado, Montana, North Dakota, South Dakota, Utah, Wyoming)
   United States Department of Labor, ETA
   Office of Job Corps
   Federal Office Building, Room 1680
   1961 Stout Street
   Denver, CO 80202
   303-844-4807

Region 9
   (Arizona, California, Hawaii, Nevada, Guam)
   United States Department of Labor, ETA
   Office of Job Corps
   71 Stevenson Street
   Box 3768, Suite 1015
   San Francisco, CA 94119
   415-995-5458

Region 10
   (Alaska, Idaho, Oregon, Washington)
   United States Department of Labor, ETA
   Office of Job Corps
   909 First Avenue, Room 1131
   Seattle, WA 98174
   206-442-1133

BRICKLAYING (PRE-APPRENTICE) JOB CORPS STATE OFFICES

Alabama
Tuskegee Job Corps Center
Tuskegee Institute
106 Moton Hall
Tuskegee, AL 36088
205-727-8803
Contractor: Tuskegee University

Job Corps Offices

Arizona
 Phoenix Job Corps Center
 518 South Third Street
 Phoenix, AZ 85004
 602-254-5921
 Contractor: Teledyne Economic Development Company

Colorado
 Collbran Job Corps Civilian Conservation Center
 P.O. Box 307
 Collbran, CO 81624
 303-487-3576
 Contractor: United States Dept. of the Interior/Bureau of Reclamation

Florida
 Jacksonville Job Corps Center
 205 W. Third Street
 Jacksonville, FL 32206
 904-353-5904
 Contractor: Teledyne Economic Development Company

Georgia
 Turner Job Corps Center
 1604 Elmer Darosa Avenue
 Albany, GA 31708
 912-883-8500
 Contractor: Management and Training Corporation

Illinois
 Golconda Job Corps Civilian Conservation Center
 Route 4, Box 104A
 Golconda, IL 62938
 618-285-6601
 Contractor: United States Dept. of Agriculture

Iowa
 Denison Job Corps Center
 Highway 30 East
 Denison, IA 51442
 712-263-4192
 Contractor: Management and Training Corporation

Kentucky
 Carl C. Perkins Job Corps Center
 Box G-11, Gobel Roberts Road
 Prestonburg, KY 41653
 606-886-1037
 Contractor: Career Systems and Development Corporation

 Earle C. Clements Job Corps Center
 Highway 60
 Morganfield, KY 42437
 502-389-2419
 Contractor, Career Systems and Development Corporation

 Great Onyx Job Corps Civilian Conservation Center
 HC 61 - Box 341
 Mammoth Cave, KY 42259
 502-286-4514
 Contractor: United States Dept. of Interior/Park Service

 Whitney Young Job Corps Center
 P.O. Box 307
 Simpsonville, KY 40067
 502-722-8862
 Contractor: RCA/GE

Maryland
 Chesapeake Job Corps Center
 Port Deposit, MD 21904
 301-939-0450
 Contractor: RCA/GE

 Woodstock Job Corps Center
 P.O. Box 8
 Wocdstock, ME 21163
 301-461-1100
 Contractor: RCA/GE

Mississippi
Gulfport Job Corps Center
3300 20th Street
Gulfport, MS 39501
601-864-9691
Contractor: Res-Care, Inc.

Mississippi Job Corps Center
P.O. Box 817
Crystal Springs, MS 39059
601-892-3348
Contractor: Res-Care, Inc.

Montana
Anaconda Job Corps Civilian
  Conservation Center
1407 Foster Creek Road
Anaconda, MT 59711
406-563-3476
Contractor: United States Dept.
  of Agriculture

New Jersey
Edison Job Corps Center
500 Plainfield Avenue
Edison, NJ 08817
201-985-4810
Contractor: ITT Educational
  Services, Inc.

New Mexico
Roswell Job Corps Center
P.O. Box 5970
Roswell, NM 88202
505-347-5414
Contractor: Vinnell Corp.

North Carolina
Kittrell Job Corps Center
P.O. Box 278, Kittrell C
Kittrell, NC 27544
919-438-6161
Contractor: Management and
  Training Corporation

Schenck Job Corps Civilian
  Conservation Center
P.O. Box 98
Pisgah Forest, NC 28768
704-877-3291
Contractor: United States Dept.
  of Agriculture

Oklahoma
Talking Leaves Job
  Corps Center
P.O. Box 948
Tahlequah, OK 74465
918-456-9959
Contractor: Cherokee Nation of
  Oklahoma

South Dakota
Boxelder Job Corps Civilian
  Conservation Center
P.O. Box 47
Nemo, SD 57759
605-348-3636
Contractor: United States Dept.
  of Agriculture

Tennessee
Jacobs Creek Job Corps Civilian
  Conservation Center
Route 1, Drawer W
Bristol, TN 37822
615-878-4021
Contractor: United States Dept.
  of Agriculture

Texas
Gary Job Corps Center
P.O. Box 967
San Marcos, TX 78667
512-396-6652
Contractor: Texas Educational
  Foundation

Utah
Weber Basin Job Corps Civilian
 Conservation Center
P.O. Box 307
Ogden, UT 84403
801-479-9806
Contractor: United States Dept.
 of Interior/Bureau of
 Reclamation

Virginia
Old Dominion Job Corps Center
P.O. Box 278
Monroe, VA 24574
804-929-4081
Contractor: Teledyne Economic
 Development Company

Washington
Columbia Basin Job Corps
 Civilian Conservation Center
Building 2402, 24th Street
Moses Lake, WA 98837
509-762-5581
Contractor: United States Dept.
 of Interior/Bureau of
 Reclamation

Curlew Job Corps Civilian
 Conservation Center
3090-100 Bamber Creek Road
Wauconda, WA 98859
509-779-2611
Contractor: United States Dept.
 of Agriculture

Wisconsin
Blackwell Job Corps Center
Route 1
Laona, WI 54541
715-674-2311
Contractor: United States Dept.
 of Agriculture

BRICKLAYING (APPRENTICE) JOB CORPS STATE OFFICES

Arkansas
Cass Job Corps Civilian
 Conservation Center
Ozark, AR 72949
501-667-3686
Contractor: United States Dept.
 of Agriculture

Ouachita Job Corps Civilian
 Conservation Center
Route 1
Royal, AR 71968
501-767-2707
Contractor: United States Dept.
 of Agriculture

California
San Diego Job Corps Center
1325 Iris Avenue,
 Building 60
Imperial Beach, CA 92032
619-429-8500
Contractor: Career Systems
 Development Corporation

District of Columbia
Potomac Job Corps Center
No. 1 DC Village Lane, SW
Washington, D.C. 20032
202-574-5000
Contractor: RCA/GE

Idaho
Marsing Job Corps Civilian
 Conservation Center
Route 1
Marsing, ID 83639
208-896-4126
Contractor: United States Dept.
 of Interior/Bureau of
 Reclamation

Illinois
Golconda Job Corps Civilian
 Conservation Center
Route 4, Box 104A
Golconda, IL 62938
618-285-6601
Contractor: United States Dept.
 of Agriculture

Indiana
Atterbury Job Corps Center
Box 187
Edinburg, IN 46124
812-526-5581
Contractor: Res-Care, Inc.

Kentucky
Frenchburg Job Corps Civilian
 Conservation Center
Box 935
Mariba, KY 40345
606-768-2111
Contractor: United States Dept.
 of Agriculture

Pine Knot Job Corps Civilian
 Conservation Center
Pine Knot, NY 42635
606-354-2176
Contractor: United States Dept.
 of Agriculture

Missouri
Mingo Job Corps Civilian
 Conservation Center
Route 2
Puxico, MO 63690
314-222-3537
Contractor: United States Dept.
 of Interior/Fish and Wildlife

Nevada
Sierra Nevada Job Corps Center
5005 Echo Avenue
Reno, NV 89506
702-677-3610
Contractor: University of
 Nevada-Reno and
 Management and Training
 Corporation

New York
Iroquois Job Corps Civilian
 Conservation Center
RR 1 Tibbits Road
Medina, NY 14103
716-798-3300
Contractor: United States Dept.
 of Interior/Fish and Wildlife

North Carolina
Lyndon B. Johnson Job Corps
 Center
466 Job Corps Drive
Franklin, NC 28734
704-524-4446
Contractor: United States Dept.
 of Agriculture

Oconaluftee Job Corps Center
Cherokee, NC 28719
704-497-5411
Contractor: United States Dept.
 of Interior/Park Service

Oklahoma
Treasure Lake Job Corps Center
Route 2, Box 30
Indiahoma, OK 73552
405-246-3203
Contractor: United States Dept.
 of Interior/Fish and Wildlife

Oregon
Angell Job Corps Center
335 NE Blogett Road
Yachats, OR 97499
503-547-3137
Contractor: United States Dept.
of Agriculture

Pennsylvania
Keystone Job Corps Center
P.O. Box 37
Drums, PA 18222
717-788-1164
Contractor: RCA/GE

Red Rock Job Corps Center
P.O. Box 218
Lopez, PA 18628
717-477-2221
Contractor: Management and
Training Corporation

Texas
McKinney Job Corps Center
Box 750
McKinney, TX 75069
214-542-2623
Contractor: Texas Educational
Foundation

Virginia
Flatwoods Job Corps Center
Route 1, Box 211
Coeburn, VA 24230
703-395-3384
Contractor: United States Dept.
of Agriculture

Washington
Fort Simcoe Job Corps
Civilian Conservation
Center
Route 1
White Swan, WA 98952
509-874-2244
Contractor: United States Dept.
of Interior/Bureau of
Reclamation

West Virginia
Harpers Ferry Job
Corps Center
P.O. Box 237
Harpers Ferry, WV 25425
304-725-2011
Contractor: United States Dept.
of Interior/Park Service

CEMENT MASONRY (PRE-APPRENTICE) JOB CORPS STATE OFFICES

Arkansas
Cass Job Corps Civilian
Conservation Center
Ozark, AR 72949
501-667-3686
Contractor: United States Dept.
of Agriculture

California
San Diego Job Corps Center
1325 Iris Avenue, Building 60
Imperial Beach, CA 92032
619-429-8500
Contractor: Career Systems
Development Corporation

Colorado
Collbran Job Corps Civilian
Conservation Center
P.O. Box 307
Collbran, CO 81624
303-487-3576
Contractor: United States Dept.
of the Interior/Bureau of
Reclamation

District of Columbia
Potomac Job Corps Center
No. 1 DC Village Lane, SW
Washington, D.C. 20032
202-574-5000
Contractor: RCA/GE

Florida
Jacksonville Job Corps Center
205 West Third Street
Jacksonville, FL 32206
904-353-5904
Contractor: Teledyne Economic
Development Company

Hawaii
Hawaii Job Corps Center
77600 Koko Head Park
Honolulu, HI 96825
808-396-1244
Contractor: Pacific Education
Foundation and Management
and Training Corporation

Illinois
Golconda Job Corps
Civilian Conservation
Center
Route 4, Box 104A
Golconda, IL 62938
618-285-6601
Contractor: United States Dept.
of Agriculture

Kentucky
Great Onyx Job Corps Civilian
Conservation Center
HC 61 - Box 341
Mammoth Cave, KY 42259
502-286-4514
Contractor: United States Dept.
of Interior/Park Service

Pine Knot Job Corps Civilian
Conservation Center
Pine Knot, KY 42635
606-354-2176
Contractor: United States Dept.
of Agriculture

Louisiana
Shreveport Job Corps Center
2815 Lillian Street
Shreveport, LA 71109
318-227-9331
Contractor: Management and
Training Corporation

Massachusetts
Grafton Job Corps Center
P.O. Box 447, Route 30
North Grafton, MA 01536
617-839-9529
Contractor: Training &
Development Corporation

Missouri
Excelsior Springs Job Corps Center
701 St. Louis Avenue
Excelsior Springs, MO 64024
816-637-5501
Contractor: MINACT, Inc.

St. Louis Job Corps Center
4333 Goodfellow Boulevard
St. Louis, MO 63120
314-679-6200
Contractor: MINACT, Inc.

New Jersey
 Edison Job Corps Center
 500 Plainfield Avenue
 Edison, NJ 08817
 201-985-4810
 Contractor: ITT Educational Services, Inc.

North Carolina
 Lyndon B. Johnson Job Corps Center
 466 Job Corps Drive
 Franklin, NC 28734
 704-524-4446
 Contractor: United States Dept. of Agriculture

Oregon
 Tongue Point Job Corps Center
 U.S. Highway 30
 Astoria, OR 97103
 503-325-2131
 Contractor: RCA/GE

 Wolf Creek Job Corps Civilian Conservation Center
 Little River Road
 Glide, OR 97443
 503-496-3507
 Contractor: United States Dept. of Agriculture

Puerto Rico
 Barranquitas Job Corps Center
 State Road #152
 Barranquitas, PR 00618
 809-857-9720
 Contractor: Puerto Rico Volunteer Youth Corps

 Ramey Job Corps Center
 P.O. Box 458
 Ramey, PR 00604
 809-890-2030
 Contractor: Puerto Rico Volunteer Youth Corps

South Carolina
 Bamberg Job Corps Center
 P.O. Box 967
 Bamberg, SC 29003
 803-245-5101
 Contractor: MINACT, Inc.

Utah
 Weber Basin Job Corps Civilian Conservation Center
 P.O. Box 307
 Ogden, UT 84403
 801-479-9806
 Contractor: United States Dept. of Interior/Bureau of Reclamation

Washington
 Cascades Job Corps Center
 830 Fruitdale Road
 Sedro Woolley, WA 98284
 206-856-3400
 Contractor: Management and Training Corporation

CEMENT MASONRY (PRE-APPRENTICE) JOB CORPS STATE OFFICES

Arizona
 Phoenix Job Corps Center
 518 South Third Street
 Phoenix, AZ 85004
 602-254-5921
 Contractor: Teledyne Economic Development Company

Arkansas
Ouachita Job Corps Civilian
  Conservation Center
Route 1
Royal, AR 71968
501-767-2707
Contractor: United States Dept.
  of Agriculture

California
Sacramento Job Corps Center
3100 Meadowview Road
Sacramento, CA 95832
916-393-2880
Contractor: Career Systems
  Development Corporation

Georgia
Brunswick Job Corps Center
Glenco Industrial Park
Brunswick, GA 31520
912-264-8843

Idaho
Marsing Job Corps Civilian
  Conservation Center
Route 1
Marsing, ID 83639
208-896-4126
Contractor: United States Dept.
  of Interior/Bureau of
  Reclamation

Indiana
Atterbury Job Corps Center
Box 187
Edinburg, IN 46124
812-526-5581
Contractor: Res-Care, Inc.

Kentucky
Frenchburg Job Corps Civilian
  Conservation Center
Box 935
Mariba, KY 40345
606-768-2111
Contractor: United States Dept.
  of Agriculture

Maryland
Woodstock Job Corps Center
P.O. Box 8
Woodstock, ME 21163
301-461-1100
Contractor: RCA/GE

Montana
Trapper Creek Job Corps
  Civilian Conservation Center
Darby, MT 59829
406-821-3286
Contractor: United States Dept.
  of Agriculture

Nebraska
Pine Ridge Job Corps Civilian
  Conservation Center
Star Route 1, Box 39-F
Chadron, NE 69337
308-432-3316
Contractor: United States Dept.
  of Agriculture

Nevada
Sierra Nevada Job Corps
  Center
5005 Echo Avenue
Reno, NV 89506
702-677-3610
Contractor: University of
  Nevada-Reno and
  Management and Training
  Corporation

North Carolina
Oconaluftee Job Corps Center
Cherokee, NC 28719
704-497-5411
Contractor: United States Dept.
  of Interior/Park Service

Oklahoma
Treasure Lake Job Corps Center
Route 2, Box 30
Indiahoma, OK 73552
405-246-3203
Contractor: United States Dept.
of Interior/Fish and Wildlife

Oregon
Timber Lake Job Corps Civilian
Conservation Center
59868 East Highway #224
Estacada, OR 97023
503-834-2291
Contractor: United States Dept.
of Agriculture

Pennsylvania
Keystone Job Corps Center
P.O. Box 37
Drums, PA 18222
717-788-1164
Contractor: RCA/GE

Red Rock Job Corps Center
P.O. Box 218
Lopez, PA 18628
717-477-2221
Contractor: Management and
Training Corporation

Tennessee
Jacobs Creek Job Corps
Civilian Conservation
Center
Route 1, Drawer W
Bristol, TN 37822
615-878-4021
Contractor: United States Dept.
of Agriculture

Texas
McKinney Job Corps Center
Box 750
McKinney, TX 75069
214-542-2623
Contractor: Texas Educational
Foundation

Virginia
Flatwoods Job Corps Center
Route 1, Box 211
Coeburn, VA 24230
703-395-3384
Contractor: United States Dept.
of Agriculture

Washington
Columbia Basin Job Corps
Civilian Conservation Center
Building 2402, 24th Street
Moses Lake, WA 98837
509-762-5581
Contractor: United States Dept.
of Interior/Bureau of
Reclamation

West Virginia
Harpers Ferry Job Corps Center
P.O. Box 237
Harpers Ferry, WV 25425
304-725-2011
Contractor: United States Dept.
of Interior/Park Service

APPENDIX E

INDUSTRY ASSOCIATIONS

From this following extensive list, you can see that the masonry industry is well supported by many organizations and associations. Talk to the association people in your state or write a national organization for more information on masonry career opportunities. They have an interest in helping the industry to grow.

NATIONAL ASSOCIATIONS

American Ceramic Society
757 Brooksedge Plaza Drive
Westerville, OH 43081
614-890-4700
W. Paul Holbrook,
   Executive Director

American Concrete Institute
   (ACI)
P.O. Box 19150
Detroit, MI 48219
313-532-2600
An association that does research in concrete construction and supports activities of the industry's engineers and contractors.

American Concrete Pavement
   Association
3800 North Wilke Road, Suite 490
Arlington Heights, IL 60004
708-394-5577
Marlin J. Knutson, President

American Institute of Architects
1735 New York Avenue, NW
Washington, DC 20006
202-626-7300
James P. Cramer, Executive
   Vice President/CEO

American Institute of Constructors
9887 North Gandy, Suite 104
St. Petersburg, FL 33702
813-578-0317
Cheryl P. Harris, Executive
   Director

Industry Associations 127

American National Standards
   Institute
11 West 42nd Street
New York, NY 10036
212-642-4900
Manual Peralta, President

American Rental Association
1900 Nineteenth Street
Moline, IL 61265
309-764-1533
James R. Irish, Executive Vice
   President

American Society for Concrete
   Construction
1902 Techny Court
Northbrook, IL 60062
708-291-0270
W. Burr Bennett, Jr., Executive
   Vice President

American Society of Professional
   Estimators
11141 George Avenue, Suite 412
Wheaton, MD 20902
302-929-8848
Beverly S. Perrell, Executive
   Director

American Subcontractors
   Association (ASA)
1004 Duke Street
Alexandria, VA 22314
703-684-3450
Chris S. Stinebert, Executive
   Vice President

Associated Builders and
   Contractors (ABC)
729 Fifteenth Street, NW
Washington, D.C. 20005
202-637-8800
Daniel J. Bennet, Executive
   Vice President
Members are primarily open
   shop (nonunion) contractors.
   ABC has more than forty
   chapters in the United
   States.

Associated Construction
   Distributors International Inc.
1505 Johnson Ferry Road,
   Suite 150
Marietta, GA 30062
404-971-2342
Daniel A. Guntin, Executive
   Vice President

Associated General Contractors of
   America (AGC)
1957 E Street, NW
Washington, DC 20006
202-393-2040
Hubert Beatty, Executive Vice
   President
AGC embraces general
   contractors in all types of
   construction and has more
   than 100 chapters in the
   United States.

Association of Specialists in
   Cleaning and Restoration
10830 Annapolis Junction Road,
   Suite 312
Annapolis Junction, MD 20701
301-604-4411

ASTM
1916 Race Street
Philadelphia, PA 19103
215-299-5400
Joseph G. O'Grady, President

BAC Job Information Center
801-5 North Second Street
St. Louis, MO 63102
314-421-5757

Brick Institute of America (BIA)
11490 Commerce Park Drive
Reston, VA 22091
703-620-0010
Nelson J. Cooney, President
BIA has a video that it produced with the NCMA that explains how a young person should consider bricklaying as a career. It is called "Bricklaying—Master the Craft and Make Your Mark." It costs thirty-five dollars.

Building Officials and Code Administrators International Inc. (BOCA)
4061 West Flossmoor Road
Country Club Hills, IL 60478
708-799-2300
Paul K. Heilstedt, Executive Director

Building Stone Institute (BSI)
P.O. Box 5047
White Plains, NY 10602
914-232-5725
Dorothy Kender, Executive Director

Cast Stone Institute
Pavilions at Greentree, Suite 408
Highway 73
Marlton, NJ 08053
609-858-0271
W. N. Russell III, Executive Director

Chimney Safety Institute of America
P.O. Box 429
Olney, MD 20832
301-774-5600
John E. Bittner, Executive Director

Clay Flue Lining Institute
P.O. Box 465
Malvern, OH 44644
216-863-0111

Concrete and Masonry Industry Firesafety Committee
5420 Old Orchard Road
Skokie, IL 60077
708-966-6200, ext. 351
James P. Hurst, Secretary

Concrete Foundation Association
P.O. Box 34745
North Kansas City, MO 64116
816-471-6686

Concrete Paver Institute (CPI)
2302 Horse Pen Road
Herndon, VA 22071
703-713-1900
David R. Smith, Director

Concrete Reinforcing Steel Institute
933 North Plum Grove Road
Schaumburg, IL 60173
708-517-1200

Concrete Sawing and Drilling Association (CSDA)
6077 Roswell Road, Suite 205
Atlanta, GA 30328
404-257-1177
Edward R. Thorn, Executive Director

Construction Industry Manufacturers Association
111 East Wisconsin Avenue, Suite 940
Milwaukee, WI 53202
414-272-0943
Fred J. Broad, President

The Construction Specifications
Institute (CSI)
601 Madison Street
Alexandria, VA 22314
703-684-0300
Joseph A. Gascoigne, Executive
Director

Council for Masonry Research
5420 Old Orchard Road
Skokie, IL 60077
708-966-6200
John P. Gleason, Jr., President

Gunite/Shotcrete Contractors
Association
P.O. Box 22738
Sylmar, CA 91342
818-896-9199
Anthony Federico, President

Expanded Shale, Clay and Slate
Institute (ESCSI)
2225 East Murray Holladay
Road, Suite 102
Salt Lake City, UT 84117
801-272-7070
John P. Ries, Executive
Director

International Association of
Concrete Repair
Specialists
P.O. Box 17402
Dulles International Airport
Washington, D.C. 20041
Milt Collins, Executive
Director

International Brick Collectors
Association
1743 Lindenhall Drive
Loveland, OH 45140
513-683-4792
Peggy B. French

International Conference of
Building Officials (ICBO)
5360 South Workman Mill
Road
Whittier, CA 90601
213-699-0541
James E. Bihr, President

International Council of Employers
of Bricklayers and Allied
Craftsmen (ICEBAC)
815 15th Street, NW
Washington, DC 20005
202-783-3788
John T. Joyce, President

International Grooving & Grinding
Association
P.O. Box 1070
Skyland, NC 28776
704-684-1989
Fred A. Gray, Executive
Director

International Masonry Institute
(IMI)
823 15th Street, NW
Washington, DC 20005
202-783-3908
Ray Lackey, Executive Vice
President
Bruce Voss is the executive
director of apprenticeship
and training. This institute
is one of the main educators
for people entering the
masonry industry or
perfecting their skills.

International Union of Bricklayers
and Allied Craftsmen (BAC)
815 15th Street, NW
Washington, DC 20005
202-783-3788
John T. Joyce, President

Marble Institute of America (MIA)
33505 State Street
Farmington, MI 48335
313-476-5558
Robert Hund, Managing Director

Mason Contractors Association of America (MCAA)
1550 Spring Road, Suite 320
Oak Brook, IL 60521
708-782-6767 or 800-533-8731
George Miller, Executive Vice President

Masonry and Concrete Saw Manufacturers Institute
30200 Detroit Road
Cleveland, OH 44145
216-899-0010
Allen P. Wherry, Secretary-Treasurer

Masonry Heater Association of North America
11490 Commerce Park Drive
Reston, VA 22091
703-620-3171
J. Gregg Borchelt, Administrator

Masonry Institute of America
2550 Beverly Boulevard
Los Angeles, CA 90057
213-388-0472
James E. Amrhein, Executive Director

The Masonry Society (TMS)
2619 Spruce Street, Suite B
Boulder, CO 80302
303-939-9700
J. L. Noland, Executive Director

National Association of Brick Distributors (NABD)
212 South Henry Street
Alexandria, VA 22314
703-549-2555

National Association of Demolition Contractors
4415 West Harrison Street, Suite 246C
Hillside, IL 60152
708-449-5959
William L. Baker, Executive Director

National Association of Home Builders (NAHB)
15th and M Streets, NW
Washington, D.C. 20005
202-822-0436
Ignacio A. Cabrera, Assistant Vice President

National Association of Women in Construction
327 South Adams Street
Fort Worth, TX 76104
817-877-5551
Paula Zang, Executive Director

National Chimney Sweep Guild
P.O. Box 429
Olney, MD 20832
301-774-5600
John E. Bittner, Executive Director

National Concrete Masonry Association (NCMA)
2302 Horse Pen Road
Herndon, VA 22071
703-713-1900
John A. Heslip, President
NCMA is made up of about 300 manufacturers of concrete masonry products. These 300 manufacturers make about seventy percent of the entire production of concrete masonry units in this country.

Industry Associations 131

National Constructors
  Association
  1730 M Street,
  Suite 900
  Washington, D.C. 20036
  202-466-8880

National Lime Association
  3601 North Fairfax Drive
  Arlington, VA 22201
  703-243-5463
  Thomas L. Potter, Executive
    Director

National Precast Concrete
  Association (NPCA)
  825 East 64th Street
  Indianapolis, IN 46220
  800-428-5732
  James E. Tilford, Executive
    Vice President

National Ready Mixed Concrete
  Association
  900 Spring Street
  Silver Spring, MD 20910
  301-587-1400
  John I. Mullarky, First Vice
    President

National Stone Association
  1415 Elliot Place, NW
  Washington, D.C. 20007
  202-342-1100
  David O. McConnell,
    Director of Administrative
    Services

National Terrazzo & Mosaic
  Association Inc.
  3169 Des Plaines Avenue,
    Suite 132
  Des Plaines, IL 60018
  708-635-7744
  Edward A. Grazzini, Executive
    Director

National Tile Roofing
  Manufacturers Association
  3127 Los Feliz Boulevard
  Los Angeles, CA 90039
  213-660-4411
  Walter F. Pruter, Executive
    Vice President

Occupational Safety and Health
  Administration (OSHA)
  950 Upshur Street NW
  Washington, DC 20011
  202-576-6339

Portland Cement Association
  5420 Old Orchard Road
  Skokie, IL 60077
  708-966-6200
  John P. Gleason, Jr., President

Post-Tensioning Institute
  1717 West Northern Avenue,
    Suite 218
  Phoenix, AZ 85021
  602-870-7540

Precast/Prestressed Concrete
  Institute (PCI)
  175 West Jackson Boulevard
  Chicago, IL 60604
  312-786-0300
  Thomas Battles, President

Ready Mix Mortar and Stucco
  Association
  50 Illinois Avenue
  Cincinnati, OH 45215
  513-761-7800
  Mike Hornback, President

Scaffold Industry Association (SIA)
  14039 Sherman Way
  Van Nuys, CA 91405
  818-782-2012
  D. Victor Saleeby, Executive
    Vice President

Scaffolding, Shoring and Forming Institute Inc.
1300 Sumner Avenue
Cleveland, OH 44115
216-241-7333
John H. Addington, Managing Director

Sealant, Waterproofing & Restoration Institute
3101 Broadway, Suite 585
Kansas City, MO 64111
816-561-8230
Kenneth R. Bowman, Executive Vice President

Southern Building Code Congress International Inc. (SBC)
900 Montclair Road
Birmingham, AL 35213
205-591-1853
William J. Tangye, Chief Executive Director

StoneExpo Federation Inc.
1601 West 5th Avenue, Suite 105
Columbus, OH 43212
614-459-0840
Pennie L. Sabel, Executive Director

The Trowel Guild
2000 N. Florida Mango Road, Suite 104
West Palm Beach, FL 33409
407-683-4876
Ralph J. Nittolo, Executive Director

Tilt-Up Concrete Association (TCA)
P.O. Box 430
Horse Shoe, NC 28742
704-891-9578

Vocational Industrial Clubs of America (VICA)
P.O. Box 3000
Leesburg, VA 22075

Women Construction Owners & Executives USA
2829 Calhoun Avenue
Chattanooga, TN 37407
615-698-8851

REGIONAL, STATE, AND LOCAL ASSOCIATIONS

Alabama
Alabama Concrete Industries
Alabama Masonry Institute
660 Adams Avenue, Suite 101
Montgomery, AL 36104
205-265-0501

Arkansas
Arkansas Ready Mixed Concrete Association
523 East Capitol
Little Rock, AR 72202
501-372-7886

Arizona
Arizona Masonry Guild Inc.
5308 North 12th Street, Suite 400
Phoenix AZ 85014
602-265-5999
Doug Schlueter, President

California
Aggregates and Concrete Association
223 San Anselmo Avenue, Suite 5
San Anselmo, CA 94960
415-457-8200

Brick Institute of California,
  also: Masonry Institute
  (Northern California)
3130 La Selva, Suite 302
San Mateo, CA 94403
415-578-0894
Thomas Polizzi, Executive
  Director

Concrete Masonry Association
  of California and Nevada
6060 Sunrise Vista Drive,
  Suite 1875
Citrus Heights, CA 95610
916-722-1700
Stuart R. Beavers, Executive
  Director

Masonry Institute of Fresno
312 West Cromwell
Fresno, CA 93711
209-439-8800
Clifford A. Wheelock, Jr.,
  President

Masonry Resource of Southern
  California
1128 East 6th Street, Suite 6
Corona, CA 91719
714-279-4198
C. Richard Writer, Director

Northern California Concrete
  Association
1900 Point West Way, Suite 104
Sacramento, CA 95815
916-924-9999

Southern California Ready
  Mixed Concrete Association
Southern California Rock
  Products Association
P.O. Box 40
South Pasadena, CA 91031
818-441-3107

Western States Clay Products
  Association
3130 La Selva, Suite 302
San Mateo, CA 94403
415-578-0894
Gordon A. Knapp, Jr.,
  Executive Director

Colorado

Colorado Ready Mixed Concrete
  Association
Colorado Rock Products Association
241 West 56th Avenue, Suite 200
Denver, CO 80216
303-296-6500

Rocky Mountain Masonry Institute
1780 South Bellaire, Suite 602
Denver, CO 80222
303-691-2141
Kelly Vigil, Executive Director

Connecticut

Connecticut Construction
  Industries Association Inc.
Connecticut Ready Mixed
  Concrete Association Inc.
912 Silas Deane Highway
Wethersfield, CT 06109
203-529-6855
Marvin B. Morganbesser,
  Executive Director

Masonry Institute of Connecticut
225 Grandview Drive
Galstonbury, CT 06033
203-659-8059
Richard Filloramo, Executive
  Director

New England Concrete Masonry
  Association
P.O. Box 30
Bloomfield, CT 06002
203-243-3977
Jacquelyn T. Coleman,
  Executive Director

District of Columbia
Masonry Institute Inc.
4853 Cordell Avenue, PH One
Bethesda, MD 20814
301-652-0115
Kenneth S. Dash, Executive Director

Florida
Florida Concrete & Products Association Inc.
649 Vassar Street
Orlando, IL 32804
407-423-8279
John F. Christensen, Jr., President

Masonry Contractors Association of Florida Inc./Central Florida Chapter
2861 Enterprise Road
DeBary, FL 32713
904-789-0670
Reg Miller, Executive Director

Georgia
Brick Institute of America Inc., Region Nine
8601 Dunwoody Place, Suite 507
Atlanta, GA 30350
404-993-9714
John C. Grogan, Executive Director

Georgia Concrete and Products Association
100 Crescent Centre Parkway, Suite 110
Tucker, GA 30084
404-621-9324
LaGrit F. Morris, Executive Director

Hawaii
Cement and Concrete Products Industry of Hawaii
2828 Paa Street, Suite 1110
Honolulu, HI 96819
808-833-1882
Steven K. L. Fong, President

Idaho
Idaho Concrete Masonry Association
1300 E. Franklin Road
Meridian, ID 83642
208-888-4050
Warren Barry, President

Illinois
Builders Association of Greater Chicago
1647 Merchandise Mart
Chicago, IL 60654
312-644-6670
Donald W. Dvorak, Executive Vice President

Concrete Contractors Association of Greater Chicago
6201 West Touhy Avenue, Suite 5
Chicago, IL 60648
312-792-0908

Illinois Concrete Council
920 South Spring
Springfield, IL 62704
217-528-1114
Randell Riley, Executive Director

Illinois Masonry Institute
Masonry Advisory Council
1480 Renaissance Drive, Suite 401
Park Ridge, IL 60068
708-297-6704
Charles Ostrander, Executive Director

Illinois Ready Mixed Concrete
  Association
229 South Rosedale
Aurora, IL 60506
708-896-6145

Mason Contractors Association
  of Greater Chicago
1480 Renaissance Drive,
  Suite 401
Park Ridge, IL 60068
708-824-0146
Jim O'Connor, Executive
  Secretary

Masonry Institute of Southern
  Illinois
Box 333
Lenzburg, IL 62255
618-475-3705
Bill Bautler, Executive
  Director

Northern Illinois Ready Mix and
  Materials Association
4415 West Harrison Street,
  Suite 435
Hillside, IL 60162
708-499-2550

Indiana
Indiana Concrete Council
Indiana Ready Mixed Concrete
  Association
9860 North Michigan Road
Carmel, IN 46032
317-872-6302

Indiana Limestone Institute of
  America Inc.
400 Stone City Bank
Bedford, IN 47421
812-275-4426
William H. McDonald,
  Executive Director

Indiana Mason Contractor
  Association
525 South Meridian Street,
  Suite 101
Indianapolis, IN 46225
317-636-9753
Terry Emmons, President

Masonry Institute of Indiana
525 South Meridian Street,
  Suite 101
Indianapolis, IN 46225
317-636-9753
David M. Sovinski, Executive
  Director

Iowa
Iowa Concrete Paving Association
Omega Place
8525 Douglas, Suite 38
Des Moines, IA 50322
515-278-0606

Iowa Ready Mixed Concrete
  Association
939 Office Park Road, Suite 201
West Des Moines, IA 50265
515-225-3535

Masonry Institute of Iowa
820 First Street, Suite 200
West Des Moines, IA 50265
515-274-9166
Larry L. Stanley, Executive
  Director

Kansas
Concrete Promotion Group Inc.
8245 Nieman Road, Suite 120
Lenexa, KS 66214
913-888-9508

Kansas Masonry Industries Council
P.O. Box 15
Ottawa, KS 66067
913-242-2177
George Mackie, Secretary

Kansas Ready Mixed Concrete
  Association
800 SW Jackson, Suite 120
Lenexa, KS 66612
913-235-1188

Missouri/Kansas Chapter,
  American Concrete Pavement
  Association
8245 Nieman Road, Suite 120
Lenexa, KS 66214
913-888-9449

Kentucky
Kentuckiana Masonry Institute
130 Fairfax Avenue, Suite 2A
Louisville, KY 40207
502-893-7840
Linda S. Muller, Executive
  Director

Kentucky Ready Mixed
  Concrete Association
136 Walnut Street
Frankfort, KY 40601
502-695-1535

Louisiana
Concrete and Aggregates
  Association of Louisiana Inc.
P.O. Box 66168
Baton Rouge, LA 70896
504-292-8692

Greater New Orleans Ready
  Mixed Concrete Association
2917 Taft Park
Metairie, LA 70002
504-455-5036

Miss-Lou Brick Manufacturers
  Association
812 North President Street
Jackson, MS 39202
601-944-1395
Kathy C. Jackson, Executive
  Director

Maryland
Masonry Institute of Maryland Inc.
1200 Cowpens Avenue
Towson, MD 21204
301-832-5855
Bernard J. Vondersmith,
  Executive Director

Maryland Ready Mix Concrete
  Association Inc.
584 Bellerive Drive, Suite 3D
Annapolis, MD 21401
301-974-4472

Massachusetts
Massachusetts Concrete and
  Aggregate Producers
  Association
P.O. Box 146, Danvers Road
Swampscott, MA 01907
617-595-0820

Masonry Institute of New England
398 Columbus Avenue, Suite 250
Boston, MA 02116
617-266-6227

Michigan
Michigan Concrete Paving
  Association
P.O. Box 19086
Lansing, MI 48901
517-371-5610

Michigan Concrete Association
2500 Kerry Street, Suite 106
Lansing, MI 48912
517-487-1181 or 800-678-9622

Mason Contractors Association
  Inc.
32080 Schoolcraft, Suite 102
Livonia, MI 48150
313-522-7350
Michelle Lind, Office
  Manager

Masonry Institute of
 Michigan Inc.
32080 Schoolcraft, Suite 102
Livonia, MI 48150
313-458-8544
Daniel S. Zechmeister,
 Executive Director

Minnesota
Brick Distributors of Minnesota
275 Market Street, Suite 409
Minneapolis, MN 55405
612-332-1545
Robert C. King, Administrative
 Manager

Minnesota Concrete & Masonry
 Contractors Association
506 Fort Road
St. Paul, MN 55102
612-293-0892
Gary Botzek, Executive Director

Minnesota Concrete Products
 Association
275 Market Street, Suite C-13
Minneapolis, MN 55405
612-332-2804
Duane Dunge, President

Minnesota Masonry Institute
275 Market Street, Suite 409
Minneapolis, MN 55405
612-332-2212
Robert C. King, Administrative
 Manager

Mississippi
Mississippi Concrete Industries
 Association
P.O. Box 12250
Jackson, MS 39236
601-936-9627
Tim Cost, Executive Director

Mississippi Concrete Paving
 Association
P.O. Box 1096
Jackson, MS 39215
601-948-3121

Miss-Lou Brick Manufacturers
 Association
812 North President Street
Jackson, MS 39202
601-944-1395
Kathy C. Jackson, Executive
 Director

Missouri
Mason Contractors Association
 of St. Louis
1429 South Big Bend
 Boulevard
St. Louis, MO 63117
314-645-1966
Wilbert Schmidt, Jr., President

Masonry Institute of St. Louis
1429 South Big Bend
 Boulevard
St. Louis, MO 63117
314-645-5888
Harry A. Fine, Executive
 Director

Missouri Concrete Association
P.O. Box 392
Jefferson City, MO 65102
314-635-6271
Randy J. Scherr, Executive
 Director

Missouri/Kansas Chapter,
 American Concrete Pavement
 Association
8245 Nieman Road, Suite 120
Lenexa, KS 66214
913-888-9449

Montana

Montana Contractors Association
Montana Ready Mix and Concrete
    Products Association Inc.
Montana Concrete Producers
P.O. Box 4519
Helena, MT 59604
406-442-4162
Ken Dunham, Secretary Manager

Nebraska

Nebraska Concrete and
    Aggregates Association
2701 North 61st Street
Lincoln, NE 68507
402-464-1897

Nebraska Concrete Council
RR 1, Box 211
Lincoln, NE 68439
402-535-2717

Nebraska Concrete Masonry
    Association
P.O. Box 7196
Omaha, NE 68107
402-330-5260
Ron Whitefoot, President

Nebraska Masonry Institute
11414 West Center Road,
    Suite 211
Omaha, NE 68144
402-330-5260
Don Littler, Executive Vice
    President

Nevada

Concrete Masonry Association
    of California and Nevada
6060 Sunrise Vista Drive,
    Suite 1875
Citrus Heights, CA 95610
916-722-1700
Stuart R. Beavers, Executive
    Director

New Jersey

New Jersey Concrete/Aggregate
    Association
770 River Road
West Trenton, NJ 08628
609-771-0099

New Jersey State Concrete
    Products Association
New Jersey Ready Mixed
    Concrete Association
7 Heather Lane
Gloversville, NY 12078
William E. Dunkinson, Jr.,
    Executive Director

New Mexico

New Mexico Concrete and
    Sand and Gravel Association
5931 Office Boulevard, NE
Albuquerque, NM 87109
505-344-7277

New York

Association of Concrete
    Contractors of Long
    Island Inc.
100 East 2nd Street
Mineola, NY 11501
516-746-7693

Capital District Masonry
    Institute
6 Airline Drive
Albany, NY 12205
518-869-0961
James G. Bradt, Managing
    Director

Concrete Industry Board
520 Fifth Avenue, 3rd Floor
New York, NY 10036
212-302-6650

Concrete Institute of New York
22 Computer Drive West
Albany, NY 12205
518-458-9233

Empire State Concrete and
    Aggregates
421 New Karner Road, Suite 10
Albany, NY 12205
518-456-0036

International Masonry Institute
P.O. Box 396
Syracuse, NY 13206
315-437-1337
John H. Harbold, Area
    Marketing Director

New York Concrete Construction
    Institute Inc.
520 Fifth Avenue
New York, NY 10036
212-302-5983

New York State Concrete
    Masonry Association Inc.
6 Airline Drive
Albany, NY 12205
518-869-3953
Gene C. Abbate, President

Precast Concrete Association of
    New York
176 York Avenue
Saratoga Springs, NY 12866
518-587-4022

North Carolina
Brick Association of North
    Carolina
324 West Wendover Avenue,
    Suite 210
Greensboro, NC 27408
919-273-5566
Peter P. Cieslak, General
    Manager

Carolinas Concrete Masonry
    Association
1 Centerview Drive, Suite 209
Greensboro, NC 27407
919-852-2074
Paul LaVene, Managing Director

The Carolinas Ready Mixed
    Concrete Association Inc.
4108 Park Road, Suite 400
Charlotte, NC 28220
704-525-2180

Carolinas`Tennessee Building
    Material Association
P.O. Box 18667
Charlotte, NC 28218
800-476-7336 or 704-376-1503
Larry Adams, Executive Vice
    President

North Carolina Masonry
    Contractors Association
Southeastern Masonry Association
3733 Benson Drive
Raleigh, NC 27609
919-872-2224
Eleanor F. Upton, Executive
    Director

North Dakota
North Dakota Ready-Mix and
    Concrete Products Association
422 North 2nd Street
Bismarck, ND 58501
701-223-2770
Curt L. Peterson, Executive
    Vice President

Ohio
Brick Institute of American -
    Mid East Region
P.O. Box 3050
North Canton, OH 44720
216-499-3001
James A. Tann, President

Masonry Institute of
  Columbus
1347 West 5th Avenue
Columbus, OH 43212
614-481-8105
Kenny Breckler, Executive
  Director

Masonry Institute of Dayton
P.O. Box 14026
Dayton, OH 45413
513-278-7821
Barbara L. Campbell, Executive
  Director

Masonry Institute of
  Northwestern Ohio
136 North Summit Avenue,
  Suite 112
Toledo, OH 43604
419-241-1912
Diane Throop, Executive
  Director

Northeastern Ohio Masonry
  Institute
21380 Lorain Road, Suite 200A
Fairview Park, OH 44126
216-333-4170
James T. Darcy, Executive
  Director

Ohio Concrete Block
  Association Inc.
17 South High Street, Suite 1200
Columbus, OH 43215
614-221-1900
Laura Alger, Executive Director

Ohio Masonry Council
P.O. Box 3050
North Canton, OH 44720
216-499-3001
James T. Darcy, Chairman

Ohio Ready Mixed Concrete
  Association
P.O. Box 29190
Columbus, OH 43229
614-891-0210

Oklahoma
Oklahoma Ready Mixed
  Concrete Associatoin
P.O. Box 3365
Edmond, OK 73083
405-341-3925

Oregon
Masonry and Ceramic Tile
  Institute of Oregon
3609 SW Corbett, Suite 4
Portland, OR 97201
503-224-1940
Larry G. Gilbertson, Executive
  Director

Oregon Concrete and Aggregate
  Producers Association Inc.
707 13th Street, SE, Suite 115
Salem, OR 97301
503-588-2430

Pennsylvania
Delaware Valley Masonry
  Institute
Meetinghouse Business Center
140 West Germantown Pike,
  Suite 240
Plymouth Meeting, PA 19462
215-940-0500
Joseph A. Barilotti, Executive
  Director

Masonry Contractors
  Association of Central
  Pennsylvania
P.O. Box 488
Dauphin, PA 17018
717-921-8642

Masonry Institute of
 Pennsylvania
2270 Noblestown Road
Pittsburgh, PA 15205
412-422-4705
Kay Lamison

Pennsylvania Aggregate and
 Concrete Association
3509 North Front Street
Harrisburg, PA 17110
717-234-2603

Pennsylvania Concrete Masonry
 Association
532 Montroyale Drive
P.O. Box 10545
Erie, PA 16514
814-825-3375
V. James Gianoni, Executive
 Director

Ready Mixed Concrete
 Association of Metropolitan
 Pittsburgh
P.O. Box 4045
Pittsburgh, PA 15201
412-364-6070

South Carolina
Brick Association of South
 Carolina
625-C Taylor Street
Columbia, SC 29201
803-252-5571
Charles E. Johnson, Executive
 Director

Carolinas-Tennessee Building
 Material Association
P.O. Box 18667
Charlotte, NC 28218
800-476-7336 or 704-376-1503
Larry Adams, Executive Vice
 President

South Dakota
South Dakota Chapter of the
 American Concrete Pavement
 Association
103 Cottonwood Drive
Spearfish, SD 57783
605-642-1523

South Dakota Ready Mixed
 Concrete Association
P.O. Box 84140
Sioux Falls, SD 57101
605-336-2928

Tennessee
Carolinas-Tennessee Building
 Material Association
P.O. Box 18667
Charlotte, NC 28218
800-476-7336 or 704-376-1503
Larry Adams, Executive Vice
 President

Masonry Institute of Tennessee
1136 Second Avenue, North
Nashville, TN 37208
615-244-3090
Darrell Lanham, Executive
 Director

Tennessee Ready Mixed
 Concrete Association
95 White Bridge Road,
 Suite 401
Nashville, TN 37205
615-353-1333

Texas
Brick Institute of Texas
Texas Association of Brick
 Distributors
P.O. Box 14667
Austin, TX 78752
512-451-4668
Glen Duncan, Executive Director

Cement and Concrete Promotion
Council of Texas
c/o Beazer West
P.O. Box 190999
Dallas, TX 75219
214-754-5505

Masonry Institute of Texas
P.O. Box 34583
Houston, TX 77075
713-941-5668
Jack E. Stubbs, Trustee

Precast Concrete Manufacturers
Association of Texas
2455 NE Loop 410, Suite 125
San Antonio, TX 78217
512-637-1386

Texas Aggregates and Concrete
Association
6633 Highway 290 E. Suite 100
Austin, TX 78723
512-451-5100

Utah

Concrete Masonry
Manufacturers of Utah
1214 East Wilmington Avenue,
Suite 301
Salt Lake City, UT 84106
801-467-6763
George Grygar, Executive
Director

Vermont

Vermont Ready-Mixed Concrete
Inc.
c/o LaFarge Corp., Northeast
Cement Division
R.D. 1, Box 1051
Williamstown, VT 05679
802-433-5371

Virginia

Mason Contractors Association
of Richmond
Mason Contractors Association
of Virginia
Virginia Masonry Council
2116 Dabney Road, Suite B-1
Richmond, VA 23230
804-358-9371
Sandra Y. Williams, Executive
Secretary

Virginia Ready Mixed Concrete
Association
620 Stagecoach Road
Charlottesville, VA 22901
804-977-3716

Washington

Mason Contractors Association,
Spokane Chapter
East 102 Boone
Spokane, WA 99202
509-326-5627
Bob Fraser, Promotion Director

Masonry Institute of Washington
925 116th Avenue, NE,
Suite 201
Bellevue, WA 98004
206-453-8820
Tim Stover, Executive Director

Northwest Concrete Masonry
Association
1409 140th Place, NE, Suite 101
Bellevue, WA 98007
206-643-6335
Thomas Young, Executive
Director

Washington Aggregates and
  Concrete Association
1800 Andover Park West,
  Suite 204
Tukwila, WA 98188
206-575-3665

Washington State Conference of
  Mason Contractors
925 116th Avenue, NE,
  Suite 201
Bellevue, WA 98004
206-453-8871
William Blackstock, President

Wisconsin

Brick Distributors of Wisconsin
P.O. Box 9282
Madison, WI 53715
608-273-2580
Gary L. Bergh, President

Fox River Valley Masons
  Promotion Fund Inc.
P.O. Box 210
De Pere, WI 54115
414-337-1647
Brad Deprez,
  Secretary-Treasurer

Wisconsin Concrete Masonry
  Association
9501 South Shore Drive
Valders, WI 54245
414-773-2888
Richard H. Walter,
  Executive Director

Wisconsin Concrete Pavement
  Association
5721 Odana Road
Madison, WI 53719
608-277-1312

Wisconsin Precast Concrete
  Association Inc.
2 East Mifflin Street,
  Suite 600
Madison, WI 53703
608-255-8891

Wisconsin Ready Mixed
  Concrete Association
9415 West Forest Home Avenue
Hales Corners, WI 53130
414-529-5077

APPENDIX F

PUBLICATIONS FOR MORE INFORMATION

Bricklaying: Brick and Block Masonry, Brick Institute of America, 11490 Commerce Park Drive, Reston, VA 22091, soft cover, 350 pages, $15.00, 1988. (a textbook; excellent details of everything it means and takes to be a journeyman bricklayer)

Filloramo, Richard, *Masonry Guide and Construction Services Manual,* Masonry Institute of Connecticut, 500 Main St., Yalesville, CT 06492, three-ring binder, 270 pages, $40.00, 1986. (Section 3 lists more than 1,800 publications available from national and international associations and suppliers.)

Harris, Cyril M., editor, *Dictionary of Architecture and Construction,* McGraw-Hill Book Co., 11 West 19th St., New York, NY 10011, soft cover, 553 pages, $19.50, 1987.

The Directory of Construction Associations, Meta Data Inc., Box 319, Huntington, NY 11743, soft cover, 407 pages, $39.95, 1987.

Building Block Walls, National Concrete Masonry Association, 2302 Horse Pen Road, Herndon, VA 22071, soft cover, 170 pages, $9.95, 1988. (A how-to guide for those entering the masonry profession.)

Gerstel, David, *Running a Successful Construction Company,* The Taunton Press, P.O. Box 5506, Newton, CT 06570, soft cover, 221 pages, $27.95, 1991. (A step-by-step guide to setting up a contracting firm.)

Job Descriptions for the Construction Industry, PAS Publications, 3101 E. Eisenhower, Ste. #2, Ann Arbor, MI 48104, three-ring binder, 184 pages, $87.50, 1987.

Kreh, Richard T., Sr. *Masonry Skills,* second ed., Delmar Publishers, Inc., P.O. Box 15015, Albany, NY 12212, hard cover, $16.00; soft cover, $12.00, 328 pages, 1982. (To be used in high school or vocational masonry programs as a guide during apprenticeship programs.)

Olin, Harold B., *Construction: Principles, Materials & Methods, 5th Ed.,* Van Nostrand Reinhold, 115 Fifth Ave., New York, NY 10003, hard cover, 1,300 pages, $52.95, 1990. (This encyclopedia covers forty-six areas of building.)

Milliner, Michael S., *Contractor's Business Handbook,* R. S. Means Co., 100 Construction Plaza, Kingston, MA 02364, hard cover, 307 pages, $53.95, 1990.

Sumichrast, Michael and C. P. McMahon, *Opportunities in Building Construction Trades,* VGM Career Horizons Series, National Textbook Company, 4255 W. Touhy, Lincolnwood, IL 60646.

Training Opportunities in the Job Corps, a Directory of Job Corps Centers and Programs, free from: U.S. Department of Labor, Employment and Training Administration, Washington, D.C. 20213.

Magazine of Masonry Construction, a how-to magazine for the masonry construction industry. Published monthly since April 1989 by The Aberdeen Group, 426 South Westgate, Addison, IL 60101 (708-543-0870). (Excellent source of how-to articles and up-to-date news on masonry construction).

Concrete Construction, a how-to magazine for the concrete construction industry. Published monthly since 1956 by The Aberdeen Group, 426 South Westgate, Addison, IL 60101 (708-543-0870). (Excellent source of how-to articles and up-to-date news on concrete construction.)

APPENDIX G
INDUSTRY TRADE SHOWS

Masonry Expo (annually)

This is the largest trade show in the United States for brick and block masonry. It combines the trade shows of the Mason Contractors Association of America (MCAA) and the National Concrete Masonry Association (NCMA). For more information, contact Chris Thiel, NCMA, 2302 Horse Pen Road, Herndon, VA 22070 (703-435-4900).

Masonry Expo is held each year; it combines two events: The Concrete Industries Exposition sponsored by the NCMA and the MCAA's International Masonry Conference and Exposition. Other sponsors are BIA, ESCSI, NABD, NLA, PCA, and TMS.

Brick Show

This is the only national trade show in the United States specifically about brick. It is produced by the National Association of Brick Distributors. For more information, contact Dan Denston, NABD, 212 S. Henry Street, Alexandria, VA 22314 (703-549-2555).

Carolina's Forum for Brick Distributors

This brick trade show has several technical presentations. Contact BANC, P.O. Box 13290, Greensboro, NC 27415 (919-273-5566).

StonExpo

Held every two years, with participation of major building stone associations. For more information, contact, The StonExpo Federation Inc., 1601 W. Fifth Avenue, #105, Columbus, OH 43212 (614-459-3904).

Marble Institute of America Exposition

Held every year in October. Contact MIA (Marble Institute of America), 33505 State Street, Farmington, MI 48024 (313-476-5558).

World of Concrete Exposition and Trade Show

Held every year in January or February. More than 900 exhibitors and 15,000 people attend this trade show that specializes in meeting the needs of concrete contractors, engineers, precasters, and ready mixed concrete manufacturers. It has more than 50 seminars for anyone in the concrete industry. For more information, contact The Aberdeen Group, 426 South Westgate, Addison, IL 60101 (708-543-0870).

National Chimney Sweep Guide Convention and Trade Show

Held every spring, this show provides seminars and manufacturer exhibits pertaining to the sweep industry. For more information, contact the National Chimney Sweep Guild, P.O. Box 429, Olney, MD 20832 (301-774-5600). John E. Bittner is the executive director.

OPPORTUNITIES IN
Available in both paperback and hardbound editions
Accounting
Acting
Advertising
Aerospace
Agriculture
Airline
Animal and Pet Care
Architecture
Automotive Service
Banking
Beauty Culture
Biological Sciences
Biotechnology
Book Publishing
Broadcasting
Building Construction Trades
Business Communication
Business Management
Cable Television
Carpentry
Chemical Engineering
Chemistry
Child Care
Chiropractic Health Care
Civil Engineering
Cleaning Service
Commercial Art and Graphic Design
Computer Aided Design and Computer Aided Mfg.
Computer Maintenance
Computer Science
Counseling & Development
Crafts
Culinary
Customer Service
Dance
Data Processing
Dental Care
Direct Marketing
Drafting
Electrical Trades
Electronic and Electrical Engineering
Electronics
Energy
Engineering
Engineering Technology
Environmental
Eye Care
Fashion
Fast Food
Federal Government
Film
Financial
Fire Protection Services
Fitness
Food Services
Foreign Language
Forestry
Gerontology
Government Service
Graphic Communications
Health and Medical
High Tech
Home Economics
Hospital Administration
Hotel & Motel Management
Human Resources Management Careers
Information Systems
Insurance
Interior Design
International Business
Journalism
Laser Technology
Law

Law Enforcement and Criminal Justice
Library and Information Science
Machine Trades
Magazine Publishing
Management
Marine & Maritime
Marketing
Materials Science
Mechanical Engineering
Medical Technology
Metalworking
Microelectronics
Military
Modeling
Music
Newspaper Publishing
Nursing
Nutrition
Occupational Therapy
Office Occupations
Opticianry
Optometry
Packaging Science
Paralegal Careers
Paramedical Careers
Part-time & Summer Jobs
Performing Arts
Petroleum
Pharmacy
Photography
Physical Therapy
Physician
Plastics
Plumbing & Pipe Fitting
Podiatric Medicine
Postal Service
Printing
Property Management
Psychiatry
Psychology
Public Health
Public Relations
Purchasing
Real Estate
Recreation and Leisure
Refrigeration and Air Conditioning
Religious Service
Restaurant
Retailing
Robotics
Sales
Sales & Marketing
Secretarial
Securities
Social Science
Social Work
Speech-Language Pathology
Sports & Athletics
Sports Medicine
State and Local Government
Teaching
Technical Communications
Telecommunications
Television and Video
Theatrical Design & Production
Transportation
Travel
Trucking
Veterinary Medicine
Visual Arts
Vocational and Technical
Warehousing
Waste Management
Welding
Word Processing
Writing
Your Own Service Business

CAREERS IN Accounting; Advertising; Business; Communications; Computers; Education; Engineering; Health Care; High Tech; Law; Marketing; Medicine; Science

CAREER DIRECTORIES
Careers Encyclopedia
Dictionary of Occupational Titles
Occupational Outlook Handbook

CAREER PLANNING
Admissions Guide to Selective Business Schools
Career Planning and Development for College Students and Recent Graduates
Careers Checklists
Careers for Animal Lovers
Careers for Bookworms
Careers for Culture Lovers
Careers for Foreign Language Aficionados
Careers for Good Samaritans
Careers for Gourmets
Careers for Nature Lovers
Careers for Numbers Crunchers
Careers for Sports Nuts
Careers for Travel Buffs
Guide to Basic Resume Writing
Handbook of Business and Management Careers
Handbook of Health Care Careers
Handbook of Scientific and Technical Careers
How to Change Your Career
How to Choose the Right Career
How to Get and Keep Your First Job
How to Get into the Right Law School
How to Get People to Do Things Your Way
How to Have a Winning Job Interview
How to Land a Better Job
How to Make the Right Career Moves
How to Market Your College Degree
How to Prepare a *Curriculum Vitae*
How to Prepare for College
How to Run Your Own Home Business
How to Succeed in Collge
How to Succeed in High School
How to Write a Winning Resume
Joyce Lain Kennedy's Career Book
Planning Your Career of Tomorrow
Planning Your College Education
Planning Your Military Career
Planning Your Young Child's Education
Resumes for Advertising Careers
Resumes for College Students & Recent Graduates
Resumes for Communications Careers
Resumes for Education Careers
Resumes for High School Graduates
Resumes for High Tech Careers
Resumes for Sales and Marketing Careers
Successful Interviewing for College Seniors

SURVIVAL GUIDES
Dropping Out or Hanging In
High School Survival Guide
College Survival Guide

VGM Career Horizons
a division of *NTC Publishing Group*
4255 West Touhy Avenue
Lincolnwood, Illinois 60646-1975